Institute of Mathematical Statistics
LECTURE NOTES–MONOGRAPH SERIES
Volume 13

# Small Sample Asymptotics

**Christopher Field**
*Dalhousie University*

**Elvezio Ronchetti**
*University of Geneva*

**Institute of Mathematical Statistics**
**Hayward, California**

The production of the *IMS Lecture Notes–Monograph Series* is managed by the IMS Business Office: Jessica Utts, IMS Treasurer, and Jose L. Gonzalez, IMS Business Manager.

Library of Congress Catalog Card Number: 90-84215

International Standard Book Number 0-940600-18-8



Printed in the United States of America

# CONTENTS

# PREFACE

The aim of this monograph is to provide a coherent development of the ideas and techniques of small sample asymptotics. This term was coined by Frank Hampel and reflects our aims of obtaining asymptotic expansions which give accurate results for small sample sizes $n$, even down to $n = 1$. The central question is that of finding good approximations to the density of statistics in situations where the computation of the exact density is intractable. The techniques presented here have a broad applicability which we have tried to demonstrate in several sections of the monograph.

This monograph brings together results from a number of papers over the past 35 years. The beginning point for the authors were the fundamental papers by Daniels (1954) and Hampel (1973) in which the basic ideas of the approximation for the mean were developed. Both authors are indebted to both Professors Daniels and Hampel for their support and encouragement in putting together this monograph.

In the development and presentation of the approximations, a basic principle was that the approximations had to be computable. To that end, we have included many numerical examples by means of which the reader can judge the quality of the approximations. Our techniques provide alternatives to Edgeworth expansions on the one hand and the bootstrap on the other. The small sample approximations are simpler than Edgeworth expansion and can be thought of as a series of low order Edgeworth expansions. We obtain much better numerical accuracy than the Edgeworth and our density approximations are always positive unlike the Edgeworth. The direct approximation of the cumulative distribution function allows us to by pass the resampling of the bootstrap in the construction of confidence intervals, for instance.

Our presentation begins with results for the mean and then uses these techniques and results to obtain approximations for more interesting statistics. In all the results, there is the key idea of recentering the underlying density at the point where we want to approximate the density of the statistics and then using a normal approximation at that point. The ideas can be developed in the complex plane by using saddlepoint and steepest descent techniques and in the reals by using conjugate densities. Where possible we give a parallel development to contrast the two approaches.

The material in the book is reasonably self-contained and should be accessible to graduate students in statistics. Our hope is that the readers will be stimulated to use and develop small sample approximations in problems of interest to them.

C.A. Field — Halifax, Canada

E.M. Ronchetti — Geneva, Switzerland

April 1990

# 1. INTRODUCTION

## 1.1. MOTIVATION

Suppose we are interested in the density $f_n$ of some statistic $T_n$ $(x_1, \cdots, x_n)$, where $x_1, \cdots, x_n$ are $n$ independent identically distributed (iid) observations with the underlying density $f$ . Unless $T_n$ and/or $f$ have special forms, one cannot usually compute analytically the distribution of $T_n$.

A first alternative is to rely on asymptotic theory. Very often one can "linearize" the statistic $T_n$ and prove that the linearized statistic is equivalent to $T_n$ as $n \to \infty$, that is the difference goes to zero in probability. This leads through the central limit theorem to many asymptotic normality proofs and the resulting asymptotic distribution can be used as an approximation to the exact distribution of $T_n$. This is certainly a powerful tool from a theoretical point of view as can be seen in some good books on the subject, e.g. Bhattacharya and Rao (1976), Serfling (1980); cf. also the innovative article by Pollard (1985). But, in spite of the fact that in some complex situations one does not have any viable alternatives, very often the asymptotic distribution does not provide a good approximation unless the sample size is (very) large. Moreover, these approximations tend to be inaccurate in the tails of the distribution.

Many techniques have been devised to increase the accuracy of the approximation of the exact density $f_n$. A well known method is to use the first few terms of an *Edgeworth expansion* (cf. for instance Feller, 1971, Chapter 16). This is an expansion in powers of $n^{-1/2}$, where the leading term is the normal density. It turns out in general that the Edgeworth expansion provides a good approximation in the center of the density, but can be inaccurate in the tails where it can even become negative. Thus, the Edgeworth expansion can be unreliable for calculating tail probabilities (the values usually of interest) when the sample size is moderate to small.

In a pioneering paper, H.E. Daniels in 1954 introduced a new type of idea into statistics by applying *saddlepoint techniques* to derive a very accurate approximation to the distribution of the arithmetic mean of $x_1, \cdots, x_n$. The key idea is as follows. The density $f_n$ can be written as an integral on the complex plane by means of a Fourier transform. Since the integrand is of the form $\exp(nw(z))$, the major contribution to this integral for large n will come from a neighborhood of the saddlepoint $z_0$, a zero of $w'(z)$. By means of the *method of steepest descent*, one can then derive a complete expansion for $f_n$ with terms in powers of $n^{-1}$. Daniels (1954) also showed that this expansion is the same as that obtained using the idea of the *conjugate density* (see Esscher, 1932; Cramér, 1938; Khinchin, 1949) which can be summarized as follows. First, recenter the original underlying distribution $f$ at the point $t$ where $f_n$ is to be approximated; that is, define the conjugate (or associate) density of $f$, $h_t$. Then use the Edgeworth expansion locally at $t$ with respect to $h_t$ and transform the results back in terms of the original density $f$. Since $t$ is the mean of the conjugate density $h_t$, the Edgeworth expansion at $t$ with respect to $h_t$ is in fact an expansion in powers of $n^{-1}$ and provides a good approximation locally at that point. Roughly speaking, a higher order approximation around the center of the distribution is replaced by local low order approximations around each point. The unusual characteristic of these expansions is that the first few terms (or even just the leading term) often give very accurate approximations in the far tails of the distribution even for very small sample sizes. Besides the theoretical reasons, one empirical reason for the excellent small sample behaviour is that saddlepoint approximations are density-like objects and do not show the polynomial-like waves exhibited for instance by Edgeworth approximations.

Another approximation closely related to the saddlepoint approximation was introduced independently by F. Hampel in 1973 who aptly coined the expression *small sample asymptotics* to indicate the spirit of these techniques. His approach is based on the idea of recentering the original distribution combined with the expansion of the logarithmic derivative $f_n'/f_n$ rather than the density $f_n$ itself. Hampel argues convincingly that this is the simplest and most natural quantity to expand. A side result of this is that the normalizing constant — that is, the constant that makes the total mass equal to 1 — must be determined numerically. This proves to be an advantage since this rescaling improves the approximation. In some cases it can even be showed that the renormalization catches the term of order $n^{-1}$ leaving the approximation with a relative error of order $0(n^{-3/2})$ ; cf. Remark 3.2, section 3.3.

The aim of this monograph is to give an introduction into concepts, theory, and applications of small sample asymptotic techniques. As the title suggests, we want to include under this heading all those techniques which are similar in the spirit to those sketched above. To be a little extreme, we want to consider "*asymptotic* techniques which work well for $n = 1$" as it has been sometimes asked from a good asymptotic theory. A very simple example in this direction is Stirling's approximation to $n!$. Exhibit 1.1 shows that the *relative error* of Stirling's approximation is never greater than 4% even down to $n = 2$.

| $n$ | $n!$ | Stirling approx. | relative error (%) |
|---|---|---|---|
| 1 | 1 | 0.92 | 8.0 |
| 2 | 2 | 1.92 | 14.0 |
| 3 | 6 | 5.84 | 2.7 |
| 4 | 24 | 23.51 | 2.0 |
| 5 | 120 | 118.02 | 1.6 |

**Exhibit 1.1**

Stirling approximation ($= \sqrt{2\pi n}(n/e)^n$) to $n!$ and relative error = | exact — approx. | / exact in %.

Note that Stirling's formula is just the leading term of a Laplacian expansion of the gamma integral defining $n!$. The original approximation, that is

$$\sqrt{2\pi}((n+\frac{1}{2})/e)^{n+\frac{1}{2}},$$

derived from

$$\Delta \log x! \sim \frac{d}{dx}\log(x+\frac{1}{2})!$$

is even more accurate, cf. Daniels (1955).

Both authors were introduced into the topic via the paper by Hampel (1973) whose original idea was motivated by the application of these techniques in robust statistics. In fact, since robust procedures are constructed to be stable in a neighborhood of a fixed statistical model, their distribution theory is more complicated than that of classical procedures like least squares estimators and F-test evaluated at the normal model. In particular, it is almost impossible to compute the exact distribution of robust procedures for a finite sample size. On the other hand, the approximations based on the asymptotic distribution are often too crude to be used in practical statistical analysis. Thus, small sample asymptotics offer the tools to compute good finite sample approximations for densities, tail areas, and confidence regions based on robust statistics. The scope of there approximations is quite broad, and they have been successfully used for likelihood and conditional inference and nonparametric statistics in addition to robust statistics. Examples and computations are provided in the later chapters.

## 1.2. OUTLINE

The monograph is organized as follows.

In chapter 2 we review briefly the Edgeworth expansions. Although this monograph does not focus directly on Edgeworth expansions, they nevertheless play an important role as local approximations in small sample asymptotics. Therefore, we do not claim to cover the large amount of literature in this area but we just review in this chapter the basic results which will be used in the development of small sample asymptotic techniques. A reader who is aleady familiar with Edgeworth expansions can skip this chapter and go directly to chapter 3.

Chapter 3 introduces the basic idea behind saddlepoint approximations from two different points of view, namely through the method of steepest descent and integration on the complex plane (sections 3.2 and 3.3) and through the method of conjugate distributions (section 3.4). The technique is derived for a simple problem, namely the approximation of the distribution of the mean of n iid random variables.

Chapter 4 shows that small sample asymptotic techniques are available for general statistics. In particular, we discuss the approximation of the distribution of L-estimators (section 4.4) and multivariate M-estimators (section 4.5) for an arbitrary underlying distribution of the observations. In each case the theoretical development is accompanied by numerical examples which show the great accuracy of these approximations.

Chapter 5 emphasizes the relationship among a number of related techniques. First, Hampel's approach is discussed in detail in section 5.2. The relation between small sample asymptotics and large deviations is presented in section 5.3. Moreover, we attempt to relate the work by Durbin and Barndorff-Nielsen (see the review paper by Reid (1988) and the references thereof) in the case of sufficient and/or exponential families to the techniques discussed so far.

In chapter 6 we present tail areas approximations and the computation of confidence intervals for multiparameter problems, especially regression. A connection to the bootstrap and a nonparametric version of small sample asymptotics obtained by replacing the underlying distribution with the empirical distribution are also discussed.

Finally, chapter 7 is devoted to some miscellaneous aspects. In section 7.1 we discuss the computational issues. In fact, it is the availability of cheap computing which makes feasible the use of small sample asymptotic techniques for complex problems. A low order, simple approximation requiring non-trivial computations is carried out at a number of points and this is the type of problem ideally suited to computers. Section 7.2 presents a potential application of small sample asymptotics as a smoothing procedure leading to nonparametric density estimation. Some applications to robust statistics are developed in section 7.3. Finally, in the remaining sections we discuss the applications of these techniques to several different problems, including the considerable amount of work in the engineering literature.

# 2. EDGEWORTH EXPANSIONS

## 2.1. INTRODUCTION

Given a central limit theorem result for a statistic, one may hope to obtain an estimate for the error involved, that is the difference between the exact distribution $F_n(t)$ of the standardized statistic and the standard normal distribution $\Phi(t)$. Typically, this result for the distribution of the sum of $n$ iid random variables is known as the Berry-Esseen theorem and takes the form

$$\sup_t |F_n(t) - \Phi(t)| \leq \frac{C_0}{\sqrt{n}}, \tag{2.1}$$

where the constant $C_0$ depends on the statistic and on the underlying distribution of the observations but not on the sample size $n$. We will discuss this result for the simplest case and mention some generalizations in section 2.2.

The inequality (2.1) suggests a way to improve the approximation of $F_n$ by considering a complete ***asymptotic expansion*** of the form

$$\sum_{j=0}^{\infty} \frac{A_j(t)}{n^{j/2}}, \tag{2.2}$$

where the error incurred by using the partial sum is of the same order of magnitude as the first neglected term,
i.e.

$$|F_n(t) - \sum_{j=0}^{r} \frac{A_j(t)}{n^{j/2}}| \leq \frac{C_r(t)}{n^{(r+1)/2}}. \tag{2.3}$$

Of course, in our case $A_0(t) = \Phi(t)$ the cumulative of the standard normal distribution and $C_0(t) \equiv C_0$ is the constant given by the Berry-Esseen theorem. Note that for any fixed $n$ and $t$ (2.2) may or may not exist and that we are just using the property (2.3) of the partial sums to approximate $F_n$. These asymptotic expansions are common in numerical analysis where they are used to approximate a variety of special functions, including Bessel functions. For a good discussion of theoretical and numerical aspects, see Henrici (1977), Ch. 11. The following example presents some typical numerical aspects of these approximations.

### Example 2.1

We consider the approximation of the Binet function

$$J(z) = \log \Gamma(z) - \frac{1}{2}\log(2\pi) - (z - \frac{1}{2})\log z + z$$

through the partial sums of the asymptotic expansion

$$\sum_{j=1}^{\infty} \frac{B_{2j}}{2j(2j-1)} z^{-2j+1},$$

where $\Gamma(z)$ is the Gamma function and $B_{2j}$ are the Bernoulli numbers given by

| 2j | 2 | 4 | 6 | 8 | 10 | 12 | ... |
|---|---|---|---|---|---|---|---|
| $B_{2j}$ | 1/6 | -1/30 | 1/42 | -1/30 | 5/66 | -691/2730 | ... |

Exhibit 2.1 shows the error curves as functions of $z$.

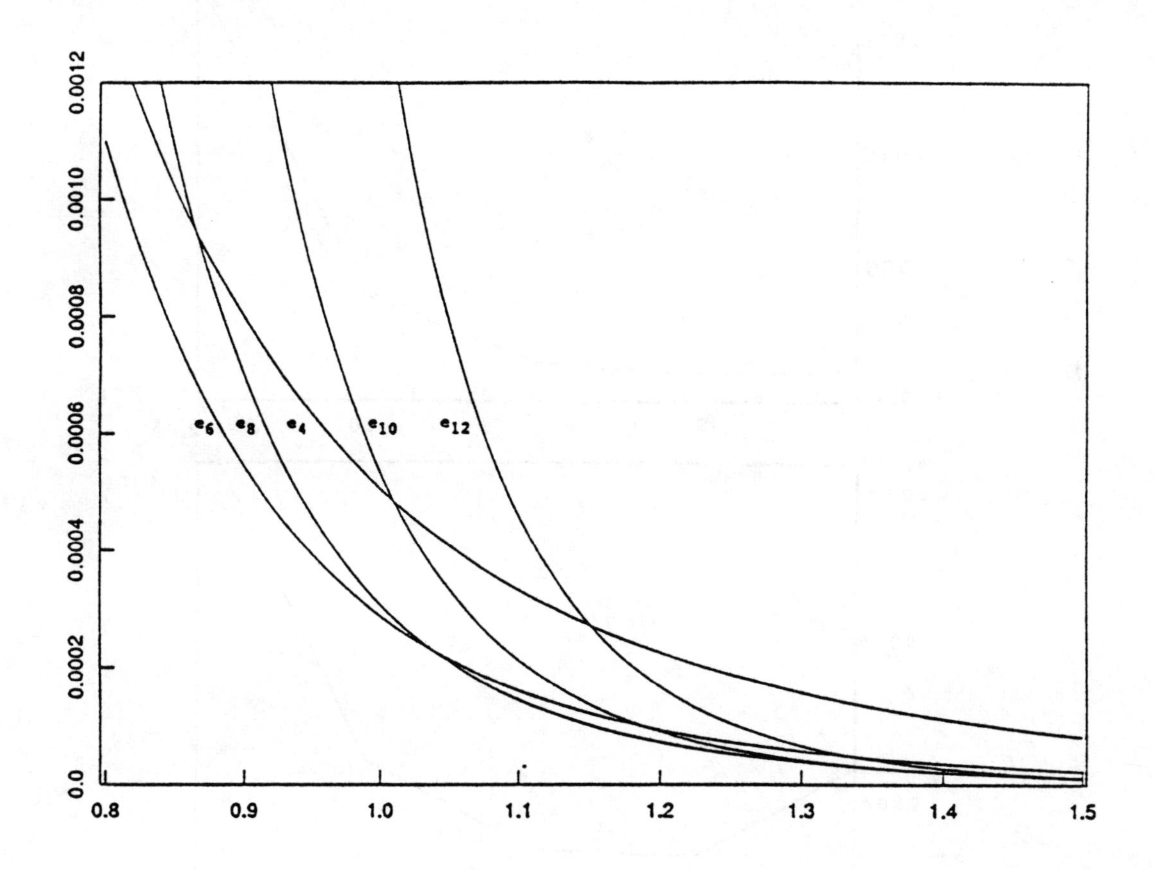


**Exhibit 2.1**

Error curves $e_r(z)$ for $r = 4, \cdots, 12$ in the approximation of the Binet function.

Exhibit 2.2 shows the error

$$e_r(z) = \left| J(z) - \sum_{j=1}^{r} \frac{B_{2j}}{2j(2j-1)} z^{-2j+1} \right|$$

as a function of the number of terms used ($r$) for various values of $z$.

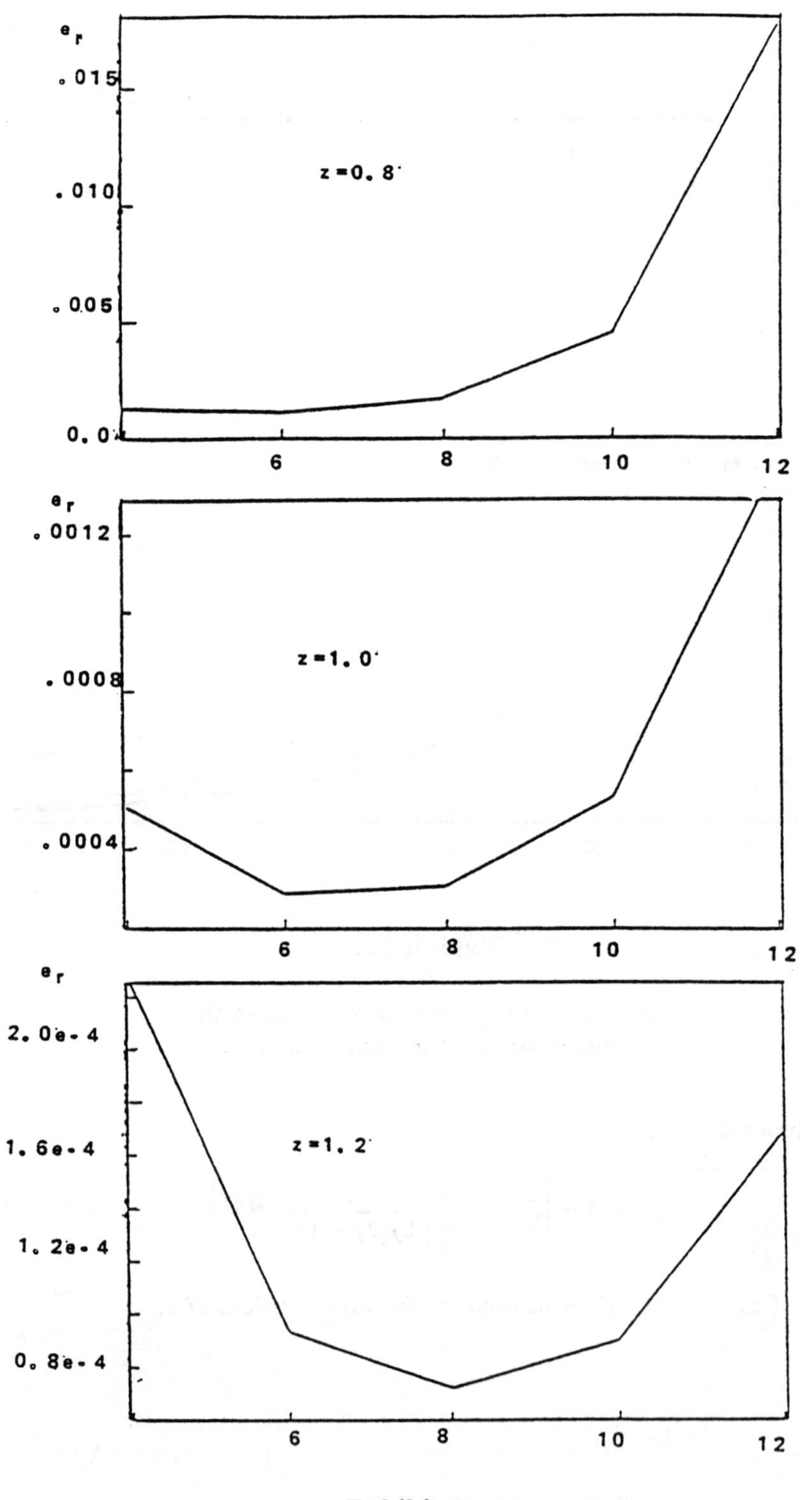


**Exhibit 2.2**

Error $e_r$ as a function of the number of terms ($r$) in the asymptotic expansion for the Binet function $J(z)$.

Both exhibits show the same story. As we add more terms to the series the error first decreases. When the number of terms $r$ reaches a certain point, adding more terms increases the error and an explosion takes place. The explosion point depends on $z$ (which plays in this example the role of $n$ in (2.2) and (2.3)) and increases as $z$ increases. Thus, in a sense as $z \to \infty$ the series has a "convergent behavior" and represents $J(z)$ asymptotically.

Edgeworth expansions are expansions of the form (2.2) with the property (2.3) holding uniformly in $t$ (i.e. $C_r$ is independent of $t$) . They play a basic role in statistical theory and practice. Since they also form the basis of one approach to saddlepoint expansions (see sections 3.4, 5.2 and chapter 4), we present in section 2.3 and 2.4 a derivation for a simple case, namely the mean of $n$ iid random variables. In sections 2.5 and 2.6 we then discuss briefly the huge amount of literature concerning Edgeworth expansions for general statistics both in univariate and multivariate situations. We conclude with some numerical examples in section 2.7.

## 2.2. CENTRAL LIMIT THEOREM AND BERRY-ESSEEN BOUND

The *central limit theorem* is the basic tool which drives asymptotic normality proofs. The resulting asymptotic distribution (the normal distribution) is used very often as an approximation to the exact distribution of some statistic $T_n$ of interest. Nowadays, central limit theorems are available for very general spaces of the observations and for very general problems. It is not our goal here to review and discuss these results that can be found in Bhattacharya and Rao (1976) and Serfling (1980). Also Pollard (1985) describes some techniques (taken from the theory of empirical processes) for proving asymptotic normality under very general conditions. In particular, the usual heavy assumptions of higher-order pointwise differentiability of the underlying density function and/or the score function defining the statistic can be dispensed with for a very broad class of statistics.

Here we want to focus on the quality of the asymptotic normal distribution as an approximation of the exact distribution. Whereas in many situations this approximation is accurate even for moderate sample sizes $n$, there are cases where it is inaccurate in the tails even for large $n$. The following example taken from Ritcey (1985) illustrates this point. The statistic of interest is

$$T_n = \frac{1}{2}\sum_{i=1}^{n}(x_i^2 + y_i^2),$$

where $x_i$ and $y_i$ are independent standard normal random variables. This test statistic is used for instance in signal detection problems where the tail probabilities of interest $P[T_n > t_0]$ are in the range $10^{-9} - 10^{-10}$ (false alarm probability). Exhibit 2.3 shows the exact probabilities (calculated from a $\frac{1}{2}\chi^2_{2n}$ distribution) and the normal approximation for various values of $n$ and $t_o$. It is clear that in this case the normal approximation is useless far out in the tails even for large $n$.

| $n$ | $t_o$ | $P$ | $\hat{P}$ | **Relative error** (%) |
|---|---|---|---|---|
| 10 | 15 | $6.98 \cdot 10^{-2}$ | $5.69 \cdot 10^{-2}$ | 18 |
| | 20 | $4.99 \cdot 10^{-3}$ | $7.83 \cdot 10^{-4}$ | 84 |
| | 25 | $2.21 \cdot 10^{-4}$ | $1.05 \cdot 10^{-6}$ | 99 |
| 100 | 125 | $9.38 \cdot 10^{-3}$ | $6.21 \cdot 10^{-3}$ | 34 |
| | 150 | $5.92 \cdot 10^{-6}$ | $2.87 \cdot 10^{-7}$ | 95 |
| | 175 | $2.78 \cdot 10^{-10}$ | $3.19 \cdot 10^{-14}$ | 99 |
| 500 | 550 | $1.15 \cdot 10^{-2}$ | $1.27 \cdot 10^{-2}$ | 10 |
| | 600 | $1.23 \cdot 10^{-5}$ | $3.87 \cdot 10^{-6}$ | 68 |
| | 625 | $1.01 \cdot 10^{-7}$ | $1.13 \cdot 10^{-8}$ | 89 |

**Exhibit 2.3**

Exact tail probabilities $P = P[T_n > t_0]$, the normal approximation $\hat{P}$, and the relative error $|P - \hat{P}|/P$ for the statistic $T_n = \frac{1}{2}\sum_{i=1}^{n}(x_i^2 + y_i^2)$, where $x_i$ and $y_i$ are independent standard normal random variables.

A first step in trying to improve the normal approximation is to assess the error involved. This is basically the Berry-Esseen bound. We first present this result for a simple case, namely the mean of $n$ iid random variables.

**Theorem 2.1** (Berry-Esseen)

Let $X_1, \cdots, X_n$ be $n$ iid random variables with distribution $F$ such that $EX_i = 0$, $EX_i^2 = \sigma^2 > 0$, $E|X_i|^3 = \rho < \infty$. Denote by $F_n$ the distribution of the standardized statistic $n^{-1/2}\sum_{i=1}^{n} X_i/\sigma$. Then, for all $n$

$$\sup_t |F_n(t) - \Phi(t)| \leq \frac{3\rho}{\sigma^3\sqrt{n}} \tag{2.4}$$

Proof: see, for instance, Feller (1971), p. 543 ff.

This result was discovered (with two different proofs) by Berry (1941) and Esseen (1942). The surprising aspect is that the Berry-Esseen bound (the right hand side of inequality (2.4)) depends only on the first three moments of the underlying distribution.

Many generalizations of this result are available today. The best known constant (which replaces 3 in (2.4)) is 0.7975, see Bhattacharya and Rao (1976), p. 110. The result can be generalized to underlying distributions $F$ without third moment and to non iid random variables. A Berry-Esseen theorem for U-statistics has been established under different sets of conditions and in increasing generality by Bickel (1974), Chan and Wierman (1977), Callaert and Janssen (1978), and Helmers and van Zwet (1982). Bjerve (1977) and Helmers (1977) proved the same result for linear combinations of order statistics. Finally, van Zwet (1984) proved a Berry-Esseen theorem for a broad class of statistics, namely symmetric functions of $n$ iid random variables.

## 2.3. CHARLIER DIFFERENTIAL SERIES AND FORMAL EDGEWORTH EXPANSIONS

The inequality (2.4) suggests a way to improve the approximation of $F_n$ by considering a complete asymptotic expansion in power of $n^{-1/2}$. This is in fact the idea of an Edgeworth expansion. Because of its conceptual simplicity, we choose in our exposition the approach via Charlier differential series as presented in Wallace (1958).

### 2.3.a Charlier Differential Series

Consider two distribution functions $H(x)$ and $G(x)$ with characteristic functions $\chi(u)$ and $\xi(u)$ and cumulants $\beta_r$ and $\gamma_r$, $r = 1, 2, \cdots$. Recall that the r-th cumulant $\beta_r$ is the r-th derivative at 0 of the cumulant generating function, i.e.

$$\beta_r = \frac{d^r}{du^r} \log \int e^{ux} dH(x)\big|_{u=0} = (-i)^r \frac{d^r}{du^r} \log \chi(u)\big|_{u=0}$$

Suppose that all derivatives of $G$ vanish at the extremes of the range of $x$. Then, by formal Taylor expansion we have

$$\log \frac{\chi(u)}{\xi(u)} = \log \chi(u) - \log \xi(u) = \sum_{r=1}^{\infty} (\beta_r - \gamma_r) \frac{(iu)^r}{r!}$$

and

$$\chi(u) = exp\left\{ \sum_{r=1}^{\infty} (\beta_r - \gamma_r) \frac{(iu)^r}{r!} \right\} \cdot \xi(u). \tag{2.5}$$

By integration by parts we can easily see that $(iu)^r \xi(u)$ is the characteristic function of $(-1)^r G^{(r)}(x)$, and by Fourier inversion of (2.5) we obtain

$$H(x) = exp\left\{ \sum_{r=1}^{\infty} (\beta_r - \gamma_r) \frac{(-D)^r}{r!} \right\} \cdot G(x), \tag{2.6}$$

where $D$ denotes the differential operator and $e^D := \sum_{j=0}^{\infty} D^j/j!$. We can now choose a distribution function $G$ which we use to develop an expansion for $H$. In fact, given a developing function $G$ and the cumulants $\beta_r$, one can formally obtain $H$ by expanding the right hand side of (2.6). Such an expansion is called Charlier differential series from Charlier (1906).

A natural developing function (but by no means the only one) is the normal distribution. In this case $G(x) = \Phi(x)$ and $(-D)^r \varphi(x) = H_r(x) \cdot \varphi(x)$, where $\varphi(x)$ is the density of the standard normal distribution and $H_r(x)$ is the Hermite polynomial of degree $r$. Chebyshev (1890) and Charlier (1905) used the normal distribution and developed (2.6) by collecting terms according to the order of derivatives. A breakthrough came from Edgeworth (1905) who applied this expansion to the distribution of the sum of $n$ iid random variables and expanded (2.6) using the normal distribution but by collecting terms according to the powers of $n$. In this way he obtained what we call today an Edgeworth expansion.

### 2.3.b Edgeworth Expansions

We now formally derive the Edgeworth expansion by applying (2.5) and (2.6) to the distribution of a standardized sum of n iid random variables. Let $X_1, \cdots, X_n$ be $n$ iid random varibles with distribution $F(x)$ and $EX_i = 0$, $varX_i = \sigma^2 > 0$, and cumulants $\kappa_r(X_i) = \lambda_r\sigma^r, r \geq 3$. Denote by $F_n(t)$ the distribution function of $T_n = n^{-1/2}\sum_{i=1}^n X_i/\sigma$ and by $\psi_n(u)$ its characteristic function. We choose the standard normal distribution as developing function, that is $G(t) = \Phi(t)$, $\xi(u) = exp(-u^2/2)$, and $\gamma_1 = 0$, $\gamma_2 = 1$, $\gamma_r = 0$, $r \geq 3$. Finally, if $\beta_r$ denotes the r-th cumulant of $T_n$ we have

$$\beta_1 = 0, \quad \beta_2 = 1,$$
$$\beta_r = \kappa_r(T_n) = n^{-r/2}\sigma^{-r}n\kappa_r(X_1) = \lambda_r n^{-(r/2-1)},$$

for $r \geq 3$. By applying (2.5) we obtain

$$\psi_n(u) = \exp\left\{\sum_{r=3}^{\infty}\frac{\lambda_r}{n^{r/2-1}}\frac{(iu)^r}{r!}\right\}e^{-u^2/2}$$
$$= \exp\left\{\frac{\lambda_3}{\sqrt{n}}\frac{(iu)^3}{3!} + \frac{\lambda_4}{n}\frac{(iu)^4}{4!} + \frac{\lambda_5}{n^{3/2}}\frac{(iu)^5}{5!} + \cdots\right\}e^{-u^2/2}$$

and by expanding $\exp\{\cdots\}$ we get

$$\psi_n(u) = \left\{1 + \frac{1}{\sqrt{n}}\lambda_3\frac{(iu)^3}{3!} + \frac{1}{n}\left[\frac{1}{2}\lambda_3^2\frac{(iu)^6}{(3!)^2} + \lambda_4\frac{(iu)^4}{4!}\right] + \frac{1}{n^{3/2}}[\cdots] + \cdots\right\}e^{-u^2/2} \tag{2.7}$$

Finally, the Fourier inversion of (2.7) leads to the following expansion

$$F_n(t) = \Phi(t) + \sum_{r=3}^{\infty}\frac{P_r(-\Phi(t))}{n^{r/2-1}}, \tag{2.8}$$

where $P_r(\cdot)$ is a polynomial of degree $3(r-2)$ with coefficients depending only on $\lambda_3, \lambda_4, \cdots, \lambda_r$. (The powers of these polynomials in (2.8) should be interpreted as derivatives.) For instance,

$$P_3(z) = \frac{\lambda_3}{6}z^3,$$
$$P_4(z) = \frac{\lambda_4}{24}z^4 + \frac{\lambda_3^2}{72}z^6.$$

For the density $f_n(t)$ one obtains the expansion

$$f_n(t) = \varphi(t) + \sum_{r=3}^{\infty}\frac{P_r(-\varphi(t))}{n^{r/2-1}}$$
$$= \varphi(t) - \frac{1}{\sqrt{n}}\frac{\lambda_3}{6}\varphi^{(3)}(t) + \frac{1}{n}\left[\frac{\lambda_4}{24}\varphi^{(4)}(t) + \frac{\lambda_3^2}{72}\varphi^{(6)}(t)\right] + \cdots$$
$$= \varphi(t) + \frac{1}{\sqrt{n}}\frac{\lambda_3}{6}H_3(t)\varphi(t) + \frac{1}{n}\left[\frac{\lambda_4}{24}H_4(t) + \frac{\lambda_3^2}{72}H_6(t)\right]\varphi(t) + \cdots, \tag{2.9}$$

where

$$H_3(t) = t^3 - 3t,$$
$$H_4(t) = t^4 - 6t^2 + 3,$$
$$H_6(t) = t^6 - 15t^4 + 45t^2 - 15$$

are the Hermite polynomials of order 3,4, and 6. (2.9) is called Edgeworth expansion for the density of $T_n$.

*Remark 2.1*
(2.9) is an expansion in powers of $n^{-1/2}$. Note however that for $t = 0$ (the expectation of the underlying distribution) all coefficients corresponding to odd powers disappear because $H_r(0) = 0$ when $r$ is odd. In this case the Edgeworth expansion becomes a series in power of $n^{-1}$:

$$f_n(0) = \frac{1}{\sqrt{2\pi}}\left\{1 + \frac{1}{n}\left[\frac{\lambda_4}{8} - \frac{5}{24}\lambda_3^2\right] + 0(n^{-2})\right\}. \tag{2.10}$$

This fact will play a key role in the derivation of the saddlepoint approximation; cf. section 3.4.

*Remark 2.2*
A similar expansion can be obtained for the $(1-\alpha)$ quantile of the distribution. This is called Fisher-Cornish expansion (see Kendall and Stuart, 1977, p. 177-179 and Cox and Hinkley, 1974, p. 464–465) and takes the form

$$q_{1-\alpha}^* + \frac{1}{\sqrt{n}}\frac{\lambda_3}{6}(q_{1-\alpha}^{*2} - 1) + \frac{1}{n}\left[\frac{\lambda_4}{24}\left(q_{1-\alpha}^{*3} - 3q_{1-\alpha}^*\right)\right.$$
$$\left. - \frac{\lambda_3^2}{36}\left(2q_{1-\alpha}^{*3} - 5q_{1-\alpha}^*\right)\right] + \cdots, \tag{2.11}$$

where $q_{1-\alpha}^*$ is the $(1-\alpha)$ quantile of the standard normal distribution, $\Phi(q_{1-\alpha}^*) = 1-\alpha$. The $1/\sqrt{n}$ and $1/n$ terms of this expansion can be interpreted as corrections to the normal quantiles taking into account skewness and kurtosis. Another method for inverting a general Edgeworth expansion is given by Hall (1983).

## 2.4. PROOF AND DISCUSSION

We present the basic idea of the proof for the density as given in Feller (1971), Ch. XVI; see also Cramer (1962).

### Theorem 2.2

Let $X_1, \cdots, X_n$ be $n$ iid random variables with common distribution $F$ and characteristic function $\psi$. Let

$$EX_i = \mu_1 = 0, varX_i = \sigma^2 < \infty,$$

and $F_n(t) = P\left[n^{-1/2}\sum_{i=1}^{n} X_i/\sigma < t\right]$ with density $f_n(t)$. Suppose that the moments $\mu_3, \cdots, \mu_k$ exist and that $|\psi|^v$ is integrable for some $v \geq 1$. Then, $f_n$ exists for $n \geq v$ and as $n \to \infty$

$$f_n(t) - \varphi(t) - \varphi(t) \sum_{r=3}^{k} P_r(t)/n^{r/2-1} = o\big(n^{-(k/2-1)}\big), \tag{2.12}$$

uniformly in $t$. Here $P_r$ is a real polynomial of degree $3(r-2)$ depending only on $\mu_1, \cdots, \mu_r$ but not on $n$ and $k$ (or otherwise on $F$).

Proof: We give an outline of the proof with the basic idea for $k = 3$. In this case we have to show that

$$f_n(t) - \varphi(t) - \frac{1}{\sqrt{n}} \frac{\mu_3}{6\sigma^3}(t^3 - 3t)\varphi(t) = o\left(\frac{1}{\sqrt{n}}\right), \tag{2.13}$$

uniformly in $t$.

Since $\psi^n(\cdot/\sqrt{n}\sigma)$ is the characteristic function of $f_n(\cdot)$, we obtain by Fourier inversion of the left hand side of (2.13)

$$\begin{aligned} &|\text{left hand side of (2.13)}| \\ &\leq \frac{1}{2\pi} \int_{-\infty}^{+\infty} \left| \psi^n\left(\frac{u}{\sqrt{n}\sigma}\right) - e^{-u^2/2} - \frac{1}{\sqrt{n}} \frac{\mu_3}{6\sigma^3}(iu)^3 e^{-u^2/2} \right| du \\ &=: N_n. \end{aligned} \tag{2.14}$$

We have to show that $N_n = o\Big(1/\sqrt{n}\Big)$ as $n \to \infty$. For a given $\delta > 0$, we split $N_n$ in two parts $N_n = N_n^{(1)} + N_n^{(2)}$, where

$$N_n^{(1)} = \frac{1}{2\pi} \int |\cdots| \cdot 1_{\{|u| > \delta\sqrt{n}\sigma\}} du,$$

and

$$N_n^{(2)} = \frac{1}{2\pi} \int |\cdots| \cdot 1_{\{|u| < \delta\sqrt{n}\sigma\}} du.$$

Since $|\psi(u)| < 1$ for $|u| \neq 0$ and $\psi(u) \to 0$ as $|u| \to \infty$, it can be shown easily that $N_n^{(1)}$ tends to 0 faster than any power of $n^{-1}$ as $n \to \infty$. The computation of $N_n^{(2)}$ requires a little more work. First rewrite $N_n^{(2)}$ as

$$\begin{aligned} N_n^{(2)} = \frac{1}{2\pi} \int e^{-u^2/2} \Bigg| &\exp\left\{ n\eta\left(\frac{u}{\sqrt{n}\sigma}\right) - 1 \right\} \\ &- \frac{1}{\sqrt{n}} \frac{\mu_3}{6\sigma^3}(iu)^3 \Bigg| \cdot 1_{\{|u| < \delta\sqrt{n}\sigma\}} du, \end{aligned} \tag{2.15}$$

where $\eta(u) = \log \psi(u) + \frac{1}{2}\sigma^2 u^2$. Then, expand $\eta(\cdot)$ in a Taylor series in a neighborhood of $u = 0$,

$$\eta(u) = \frac{\eta'''(u^*)}{6} \cdot u^3, \tag{2.16}$$

where $u^*$ is a point in the interval $]0, \delta[$. Finally, (2.16) allows us to approximate the right hand side of (2.15) to get $N_n^{(2)} = o(1/\sqrt{n})$ and this completes the proof. □

From the expansion for the density $f_n$ given in Theorem 2.2, one can obtain by integration an expansion for the cumulative distribution function $F_n$. However, this method is not available when the integrability condition on $|\psi|^\nu$ fails. In this case the following important "smoothing technique" can be used.

**Lemma** (Esseen's smoothing Lemma).

Let $H$ be a distribution with zero expectation and characteristic function $\chi$. Suppose $H - G$ vanishes at $\pm\infty$ and that $G$ has a derivative $g$ such that $|g| \leq m$. Finally, suppose that $g$ has a continuously differentiable Fourier transform $\xi$ such that $\xi(0) = 1$ and $\xi'(0) = 0$. Then,

$$|H(x) - G(x)| \leq \frac{1}{\pi} \int_{-T}^{T} \left| \frac{\chi(u) - \xi(u)}{u} \right| du + \frac{24m}{\pi T} \tag{2.17}$$

holds for all $x$ and $T > 0$.

Proof: see Feller (1971), Ch. XVI, section 3.

The following theorem establishes the Edgeworth expansion for the cumulative distribution function.

**Theorem 2.3**

Let $X_1, \cdots, X_n$ be $n$ iid random variables with common distribution $F$. Let

$$EX_i = 0, \quad varX_i = \sigma^2 < \infty, \text{ and } F_n(t) = P\left[n^{-1/2} \sum_{i=1}^{n} X_i/\sigma < t\right].$$

If $F$ is not a lattice distribution and if the third moment $\mu_3$ of $F$ exists, then

$$F_n(t) - \Phi(t) - \frac{1}{\sqrt{n}} \frac{\mu_3}{6\sigma^3}(1 - t^2)\phi(t) = o(n^{-1/2})$$

uniformly in $t$.

Proof: (Feller, 1971, Ch. XVI) Define $G(t) = \Phi(t) + 1/\sqrt{n}\mu_3/6\sigma^3(1 - t^2)\phi(t)$. Then $G$ satisfies the conditions of Esseen's Lemma with $\xi(u) = [1 + \mu_3(iu)^3/(\sqrt{n} \cdot 6\sigma^3)]e^{-u^2/2}$.

We now apply inequality (2.17) with $T = a\sqrt{n}$, where the constant $a$ is chosen such that $24|g(t)| < \epsilon \cdot a$ for all $t$ and a given $\epsilon$. Then

$$|F_n(t) - G(t)| \leq \int_{-a\sqrt{n}}^{a\sqrt{n}} \left| \frac{\psi^n(u/\sqrt{n}\sigma) - \chi(u)}{u} \right| du + \frac{\epsilon}{\sqrt{n}}. \tag{2.18}$$

Now we can apply the same arguments as in the proof of Theorem 2.2 to the integral on the right hand side of (2.18) and the result follows. □

We conclude this section with a discussion of some points on Edgeworth expansions.

1) If the underlying distribution $F$ has moments of all orders, one is tempted to let $k \to \infty$ in (2.12). Unfortunately, the resulting infinite series need not converge for any $n$. In fact Cramér showed that this series converges for all $n$ if and only if $e^{x^2/4}$ is integrable with respect to $F$; see Feller (1971), p. 542. This is in agreement with the discussion in section 2.1 In particular, adding higher order terms does not necessarily improve the approximation and can be disastrously bad; cf. Exhibits 2.1 and 2.2.
2) In section 2.3 we derived Edgeworth expansions via Charlier differential series by using the normal distribution as developing function. When the asymptotic distribution is not normal, similar expansions may be obtained by using the asymptotic distribution as developing function.
3) The approximation provided by the Edgeworth expansion is in general reliable in the center of the distribution for moderate sample sizes; see Remark 2.1, section 2.3.b and the examples in section 2.7. This makes it a suitable tool for local asymptotic theory. Unfortunately, the approximation deteriorates in the tails where it can even become negative; see section 2.7. Moreover, the absolute error is uniformly bounded over the whole range of the distribution, but the relative error is in general unbounded. This is in contrast with saddlepoint techniques which give approximations with uniformly small relative errors; cf. chapters 3, 4, and 6.

## 2.5. EDGEWORTH EXPANSIONS FOR GENERAL STATISTICS

In this section we want to discuss briefly some results on Edgeworth expansions for more complicated statistics than the arithmetic mean. Given the relative importance of U-statistics in the literature, we focus our presentation on this class of statistics.

Given a statistic $T_n$ one can in principle still use the approach via Charlier differential series presented in section 2.3.a and go through the steps (2.5) to (2.8). However, there are two points which make life hard. The first one is the fact that in general the cumulants of the statistics $\kappa_r(T_n)$ cannot be expressed in terms of $\kappa_r(X_1)$, the cumulants of the underlying distribution of the observations. The second point is that the validity of (2.8) must be proven. These problems can be overcome by replacing the exact cumulants $\kappa_r$ by approximations of order $n^{-(r/2-1)}$ and by using Esseen's smoothing Lemma to prove the validity of the resulting expansion. Most of the work lies in the estimation of the integral on the right hand side of (2.17) in order to obtain an upper bound of the appropriate order in $n$. Let us now see how this idea applies to U-statistics. Let $X_1, \cdots, X_n$ be $n$ iid random variables with common distribution $F$. Then a one-sample *U-statistic of degree 2* is defined by

$$U_n = \binom{n}{2}^{-1} \sum_{1 \le i < j < n} h(X_i, X_j), \tag{2.19}$$

where $h$ is a symmetric function of two variables with $Eh(X_1, X_2) = 0$ and $Eh^2(X_1, X_2) < \infty$.

Define

$$g(x) = E\left[h(X_1, X_2)|X_1 = x\right],$$
$$\psi(x, y) = h(x, y) - g(x) - g(y),$$
$$\hat{U}_n = \frac{1}{n}\sum_{i=1}^{n} 2g(X_i), \text{ and}$$
$$\Delta_n = \frac{2}{n(n-1)} \sum_{i \le i < j \le n} \psi(X_i, X_j).$$

Then

$$U_n = \hat{U}_n + \Delta_n. \tag{2.20}$$

Note that $2g(x) = IF(x; U, F)$, *the influence function of* $U$, defined by

$$IF(x; U, F) = \lim_{\epsilon \to 0}\left[U((1-\epsilon)F + \epsilon\Delta_x) - U(F)\right]/\epsilon,$$

where $U(\cdot)$ is the functional defined by $U(F) = E_F(U_n)$ and $\Delta_x$ is the distribution which puts mass 1 at a point $x$; see Hampel (1968, 1974), Hampel et al. (1986), and section 7.3.a. Thus, (2.20) is the linear representation of $U_n$ based on the influence function and corresponding von Mises expansion; see von Mises (1947).

U-statistics were introduced by Hoeffding (1948) who also proved the asymptotic normality. Note the important special case $h(x, y) = 1_{\{x<y\}}$ which defines the Wilcoxon statistic. Subsequently Berry-Esseen bounds for U-statistics were established by several authors; cf. section 2.2. Finally, an Edgeworth expansion of order $o(n^{-1})$ was derived by Callaert, Janssen, and Veraverbeke (1980) and more recently by Bickel, Götze, and van Zwet (1986). We will present the result of the latter paper where the assumptions appear to be very mild.

Define

$$\sigma_g^2 = Eg^2(X_1),$$
$$\lambda_3 = \sigma_g^{-3}\left\{Eg^3(X_1) + 3Eg(X_1)g(X_2)\psi(X_1, X_2)\right\},$$
$$\lambda_4 = \sigma_g^{-4}\left\{Eg^4(X_1) - 3\sigma_g^4 + 12Eg^2(X_1)g(X_2)\psi(X_1, X_2) + 12Eg(X_1)g(X_2)\psi(X_1, X_3)\psi(X_2, X_3)\right\},$$
$$\sigma_n^2 = var(U_n) = var(\hat{U}_n) + var(\Delta_n) = \frac{1}{n}4\sigma_g^2 + \frac{2}{n(n-1)}E\psi^2(X_1, X_2),$$

and

$$F_n(t) = P\left[U_n/\sigma_n < t\right].$$

Then we have the following theorem.

**Theorem 2.4** (Bickel, Götze, van Zwet, 1986).

Suppose that there exist a number $r > 2$ and an integer $k$ such that $(r-2)(k-4) > 8$ and that the following assumptions are satisfied

$$E\left|\psi(X_1,X_2)\right|^r < \infty,$$

$$E\left|g(X_1)\right|^4 < \infty,$$

$$\limsup_{|t|\to\infty} \left|E\ exp\{itg(X_1)\}\right| < 1.$$

Let $|\lambda_1| \geq |\lambda_2| \geq \cdots$ be the eigenvalues of the kernel $\psi$, that is,

$$\int \psi(x_1,x_2)w_j(x_1)dF(x_1) = \lambda_j \cdot w_j(x_2), j = 1,2,\cdots$$

and suppose that there exist $k$ nonzero eigenvalues.

Then

$$F_n(t) - \Phi(t) + \phi(t)\left\{\frac{1}{\sqrt{n}}\frac{\lambda_3}{6}(t^2-1) + \frac{1}{n}\frac{\lambda_4}{24}(t^3-3t) + \frac{1}{n}\frac{\lambda_3^2}{72}(t^5 - 10t^3 + 15t)\right\}$$

$$= o(n^{-1})$$

uniformly in $t$.

Note that $\lambda_3/\sqrt{n}$ and $\lambda_4/n$ are approximations with error $o(n^{-1})$ to the standard third and fourth cumulant of $U_n/\sigma_n$. Basically the conditions on $g(\cdot)$ in Theorem 2.4 establish an Edgeworth expansion for the *linearized statistic* $\hat{U}_n$ while the moment assumption on $\psi(\cdot,\cdot)$ allows to correct the expansion for the remainder term $\Delta_n$ in (2.20).

Results on Edgeworth expansions for other classes of statistics abound. Bickel (1974) gives a complete account of the literature up to 1974. A basic paper is Albers, Bickel, and van Zwet (1976). Results on L-statistics are discussed in Helmers (1979, 1980) and van Zwet (1979) and will be the starting point in section 4.4. More references can be found in Skovgaard (1986), p. 169-170 and Hall (1983).

## 2.6. MULTIVARIATE EDGEWORTH EXPANSIONS

The techniques used to derive Edgeworth expansions for the distribution of one - dimensional statistics can be generalized to the multivariate case. The basic idea, i.e. expansion of the characteristic function and Fourier inversion, is the same. However, in the multivariate case the notation becomes more complex. The Edgeworth expansion for the multivariate mean is discussed, among others, in Barndorff-Nielsen and Cox (1979), Skovgaard (1986), and McCullagh (1987). Explicit expressions for the multivariate Hermite polynomials appearing in the terms of the expansion are given in Grad (1949), Barndorff-Nielsen and Pedersen (1979), and Holly (1986).

A basic result on multivariate Edgeworth expansions is that of Bhattacharya and Ghosh (1978). Given a sequence of $n$ iid m-dimensional random vectors $Y_1,\cdots,Y_n$ and a sequence of $k$ real functions $f_1,\cdots,f_k$ on $\mathbf{R}^m$, Bhattacharya and Ghosh derive the rates of convergence to normality and asymptotic expansions of the distribution of statistics of the form $W_n = \sqrt{n}(H(\bar{Z}) - H(\mu))$, where $Z_i = (f_1(Y_i),\cdots,f_k(Y_i))$, $\bar{Z} = n^{-1}\sum_{i=1}^{n} Z_i$, $\mu = EZ_i$, and

$H : \mathbf{R}^k \rightarrow \mathbf{R}^p$. An important application of this result yields the asymptotic expansion of the distribution of maximum likelihood estimators. Since in sections 4.2 and 4.5 this result will be used critically to derive saddlepoint approximations to the distribution of M-estimators, we refer the reader to those sections for a detailed discussion of the conditions on the validity of the expansions.

## 2.7. EXAMPLES

In this section we discuss two examples which show the numerical aspects of the approximations based on Edgeworth expansions.

In the first one we consider the approximation of the distribution of the mean of 5 uniform $[-1, 1]$ observations. In this case the density of the mean can be computed exactly; it is given in section 3.5. In order to compare with the Edgeworth approximation, we will consider here the density of the standardized mean $\bar{X}_n/\sigma_n$, where $\sigma_n^2 = var\bar{X}_n = 1/3n$. For the density we obtain

$$f_{\bar{X}_n/\sigma_n}(t) = \frac{n^n}{\sqrt{3n}2^n(n-1)!} \sum_{i=0}^{n} (-1)^i \binom{n}{i} \left\langle 1 - \frac{t}{\sqrt{3n}} - \frac{2i}{n} \right\rangle^{n-1} \tag{2.21}$$

and for the cumulative distribution function

$$F_{\bar{X}_n/\sigma_n}(t) = \frac{n^{n-1}}{2^n(n-1)!} \sum_{i=0}^{n} (-1)^i \binom{n}{i} \left\{ \left(2 - \frac{2i}{n}\right)^n - \left(1 - m(i,t) - \frac{2i}{n}\right)^n \right\}, \tag{2.22}$$

where $|t| \leq \sqrt{3n}$, $\langle z \rangle = max(z, 0)$, and $m(i,t) = min(t/\sqrt{3n}, 1 - 2i/n)$. The corresponding approximations based on Edgeworth expansions to terms of order $n^{-1}$ are given by

$$\tilde{f}_{\bar{X}_n/\sigma_n}(t) = \left\{ 1 - \frac{1}{20n}(t^4 - 6t^2 + 3) \right\} \phi(t) \tag{2.23}$$

and

$$\tilde{F}_{\bar{X}_n/\sigma_n}(t) = \Phi(t) - \frac{1}{20n}(3t - t^3)\phi(t) \tag{2.24}$$

where $\phi(t)$ and $\Phi(t)$ are the density and the cumulative of the standard normal distribution respectively.

Exhibit 2.4 shows the error (exact-Edgeworth) in the approximation of the density whereas Exhibits 2.5 and 2.6 show the *percent relative error* for the cumulative distribution. In these exhibits, the horizontal axis is in standardized units.

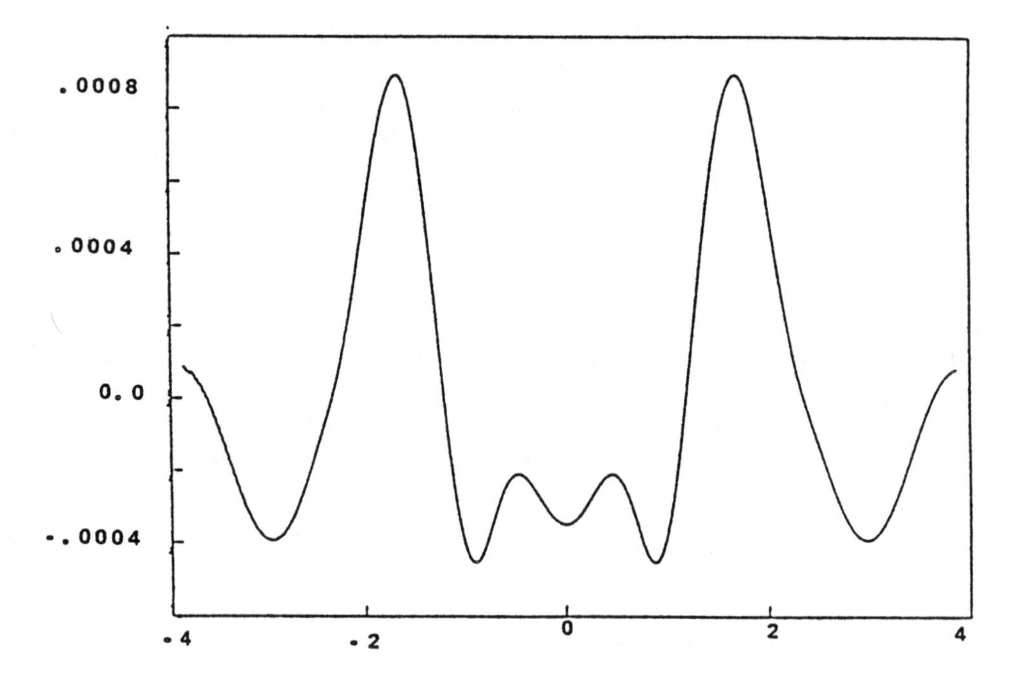


**Exhibit 2.4**
Error (exact-Edgeworth) in the approximation of the density of the mean of 5 uniform $[-1, 1]$ observations.

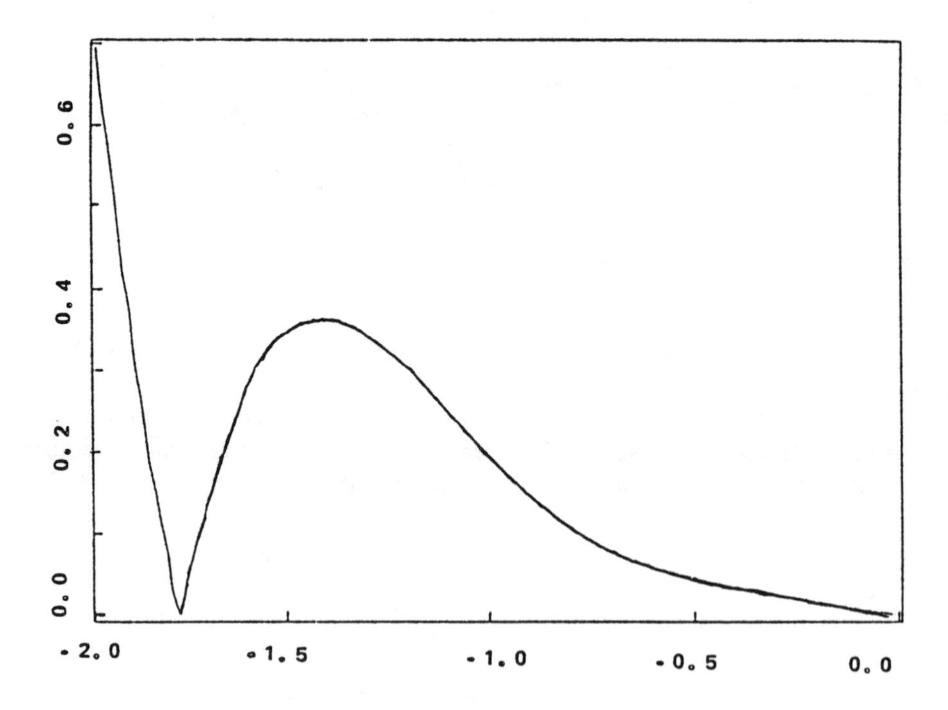


**Exhibit 2.5**
Relative error (%) in the Edgeworth approximation of the cumulative distribution function of the mean of 5 uniform $[-1, 1]$ observations.

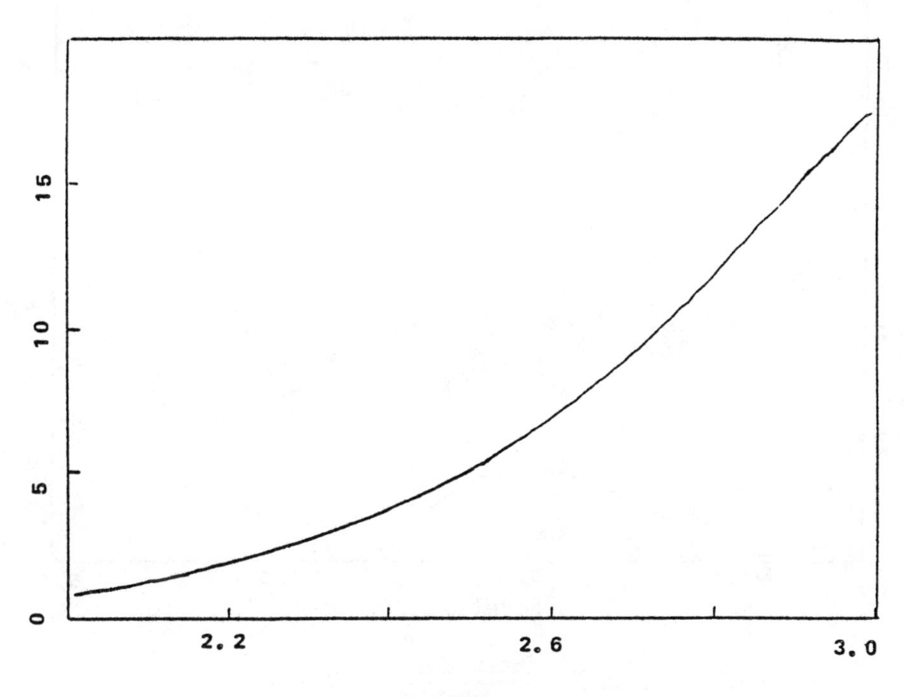


**Exhibit 2.6**

Relative error $\left(100 \times \frac{|\text{exact-Edgeworth}|}{1-\text{exact}}\right)$ in the Edgeworth approximation of the upper tail area of the distribution of the mean of 5 uniform $[-1, 1]$ observations.

While in the range $(-2, 0)$ the relative error is small, it is clear from Exhibit 2.6 that it can get up to 18% in the tail. Moreover, Exhibit 2.4 shows the typical polynomial behavior of the Edgeworth approximation. This contrasts with the uniformly small (over the whole range) relative error of the small sample asymptotic approximation, cf. Exhibits 3.7 and 3.8 and Hampel (1973). In particular, in this case the relative error is always smaller than 2.51% even in the extreme tails.

As a second example consider a Gamma distribution with shape parameter $\alpha$ (and scale parameter $\theta = 1$) as an underlying distribution:

$$f_\alpha(x) = e^{-x} x^{\alpha-1} / \Gamma(\alpha), \quad x \geq 0.$$

The density of the mean of $n$ iid observations with this underlying distribution is again a Gamma with shape parameter $n\alpha$ and scale parameter $n$. Moreover,

$$E\bar{X}_n = \alpha, \quad var\bar{X}_n = \alpha/n.$$

Exhibits 2.7 to 2.10 show the relative errors for $n = 4, 10$ and $\alpha = 2$. The same comments as in the first example apply here. Note that in this case the saddlepoint approximation is exact *(after renormalization)*; cf. Remark 3.3.

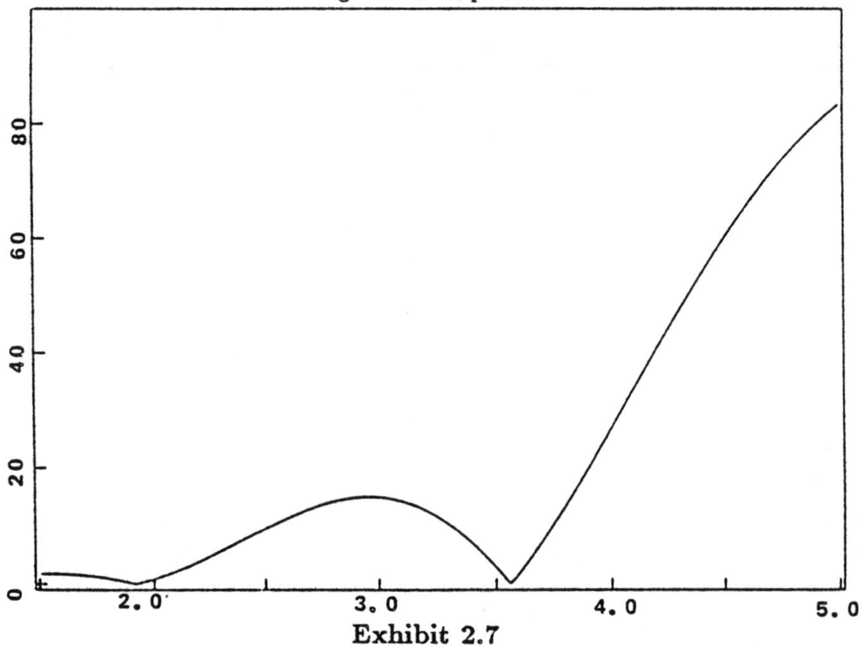


**Exhibit 2.7**

Relative error in % $\left(\frac{|\text{exact-Edgeworth}|}{1-\text{exact}} \times 100\right)$ in the Edgeworth approximation of the upper tail area of the distribution of the mean of 4 observations from a Gamma distribution with shape parameter $\alpha = 2$.

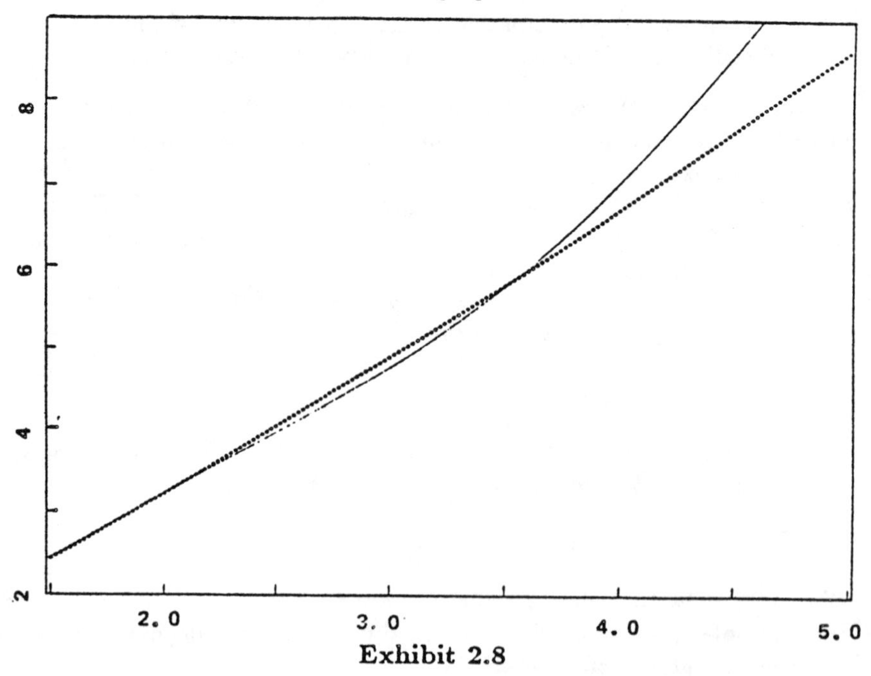


**Exhibit 2.8**

$\log(F/(1-F))$ for the exact $(\cdots)$ and the Edgeworth approximation (—) for the upper tail area for the mean of 4 observations from a Gamma distribution with shape parameter $\alpha = 2$.

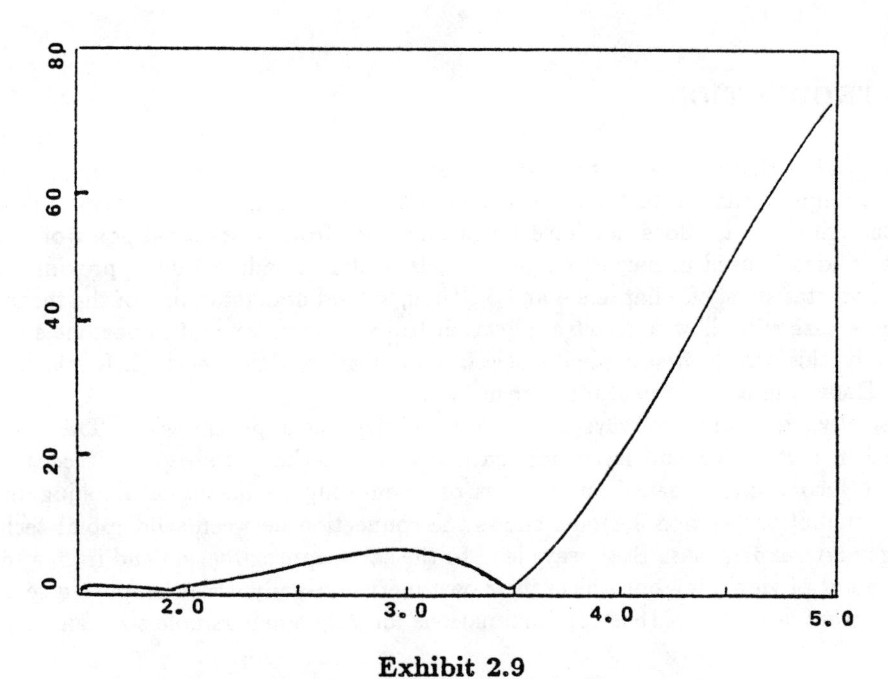


**Exhibit 2.9**
Relative error in % in the Edgeworth approximation of the upper tail area for the mean of 10 observations from a Gamma distribution with shape parameter $\alpha = 2$.

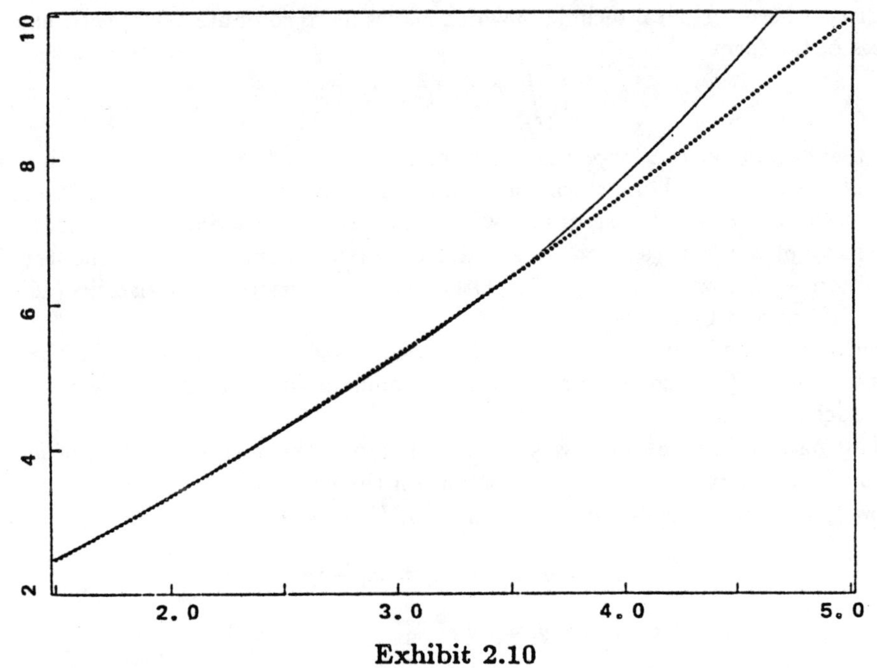


**Exhibit 2.10**
$\log(F/(1-F))$ for the exact $(\cdots)$ and the Edgeworth approximation (—) for the upper tail area of the distribution of the mean of 10 observations from a Gamma distribution with shape parameter $\alpha = 2$.

More examples can be found in sections 4.4 and 5.3.

# 3. SADDLEPOINT APPROXIMATIONS FOR THE MEAN

## 3.1. INTRODUCTION

The goal of this chapter is to introduce saddlepoint techniques for a simple problem, namely the approximation to the distribution of the mean of $n$ iid random variables.

Although this case does not have much relevance from a practical point of view, the same basic idea is used in more complex models to derive saddlepoint approximations for very general statistics; cf. chapters 4 and 5. Thus, a good understanding of the technique in this simple case will allow a direct application to more complex and important situations. Historically, this was the first explicit statistical application of this method. It was developed by H.E. Daniels in a fundamental paper in 1954.

Basically, there are two ways to derive a saddlepoint approximation. The first one is presented in section 3.3 and is an application of the method of steepest descent (section 3.2). The second one is based on the idea of recentering by means of a conjugate or associate distribution (section 3.4) and shows the connection between saddlepoint techniques and Edgeworth expansions. Both ways lead to the same approximation and from a methodological point of view they both have their own merits. Finally, the examples in section 3.5 show the great accuracy of these approximations for very small sample sizes and far out in the tails.

## 3.2. THE METHOD OF STEEPEST DESCENT

We discuss here a general technique which allows us to compute asymptotic expansions of integrals of the form

$$\int_{\mathcal{P}} e^{v \cdot w(z)} \xi(z) dz \tag{3.1}$$

when the real parameter $v$ is large and positive. Here $w$ and $\xi$ are analytic functions of $z$ in a domain of the complex plane which contains the path of integration $\mathcal{P}$. This technique is called the *method of steepest descent* and will be used to derive saddlepoint approximations to the density of a mean (section 3.3) and later of a general statistic (chapter 4). In our exposition we follow Copson (1965). Other basic references are DeBruijn (1970), and Barndorff-Neilsen and Cox (1989).

Consider first the integral (3.1). In order to compute it we can deform arbitrarily the path of integration $\mathcal{P}$ provided we remain in the domain where $w$ and $\xi$ are analytic. We deform $\mathcal{P}$ such that

(i) the new path of integration passes through a zero of the derivative $w'(z)$ of $w$;
(ii) the imaginary part of $w$, $\Im w(z)$ is constant on the new path.

Let us now look at the implications of (i) and (ii). If we write

$$z = x + iy, \qquad z_0 = x_0 + iy_0,$$

$$w(z) = u(x, y) + iv(x, y), \qquad w'(z_o) = 0,$$

and denote by $S$ the surface $(x, y) \mapsto u(x, y)$, then by the Cauchy-Riemann differential equations

$$u_x = v_y, \qquad u_y = -v_x,$$

it follows that the point $(x_0, y_0)$ cannot be a maximum or a minimum but must be a *saddlepoint* on the surface $S$. Moreover, the orthogonal trajectories to the level curves $u(x, y)$ = constant are given (again by the Cauchy-Riemann differential equations) by the curves $\Im w(z) = v(x, y)$ = constant. Since the paths on $S$ corresponding to the orthogonal trajectories of the level curves are paths of steepest (ascent) descent, condition (ii) above means that the integration along a path where $\Im w(z)$ is constant implies that we are moving along the paths of steepest descent from the saddlepoint $(x_0, y_0)$ on the surface $S$.

Therefore, on a steepest path through the saddlepoint we have

$$\begin{aligned} w(z) &= u(x, y) + iv(x_o, y_o) \\ &= u(x_0, y_0) + iv(x_0, y_0) - (u(x_0, y_0) - u(x, y)) = w(z_0) - \gamma(x, y), \end{aligned} \tag{3.2}$$

where $\gamma$ is the real function

$$\gamma(x, y) = u(x_0, y_0) - u(x, y). \tag{3.3}$$

It follows directly that $d\gamma/ds = \pm|w'(z)|$, where $s$ is the arc length of the path (on the plane). Thus, $\gamma$ is monotonic on the steepest path from the saddlepoint and either increases to $+\infty$ or decreases to $-\infty$. Since by (3.1) and (3.2) $\gamma \to -\infty$ leads to a divergent integral, we choose the path where $\gamma$ increases to $+\infty$. This is the path of steepest descent from the saddlepoint. Exhibit 3.1 shows this path for the function $w(z) = z^2$ and Exhibit 3.2 shows the surface $u = u(x, y)$ about the saddlepoint $(x_0, y_0) = (0.25, 0)$ and the path of steepest descent for the function $w(z) = K(z) - z \cdot t$, where $K(z) = -\beta \log(1 - z/\theta)$; $t > 0$, $\theta > 0$, $\beta > 0$ fixed. With this second choice of $w(z)$ and $v = n$, $\xi(z) \equiv n(2\pi i)^{-1}$, the integral (3.1) is just the density (evaluated at $t$) of the mean of $n$ iid random variables from a Gamma distribution; cf. (3.6).

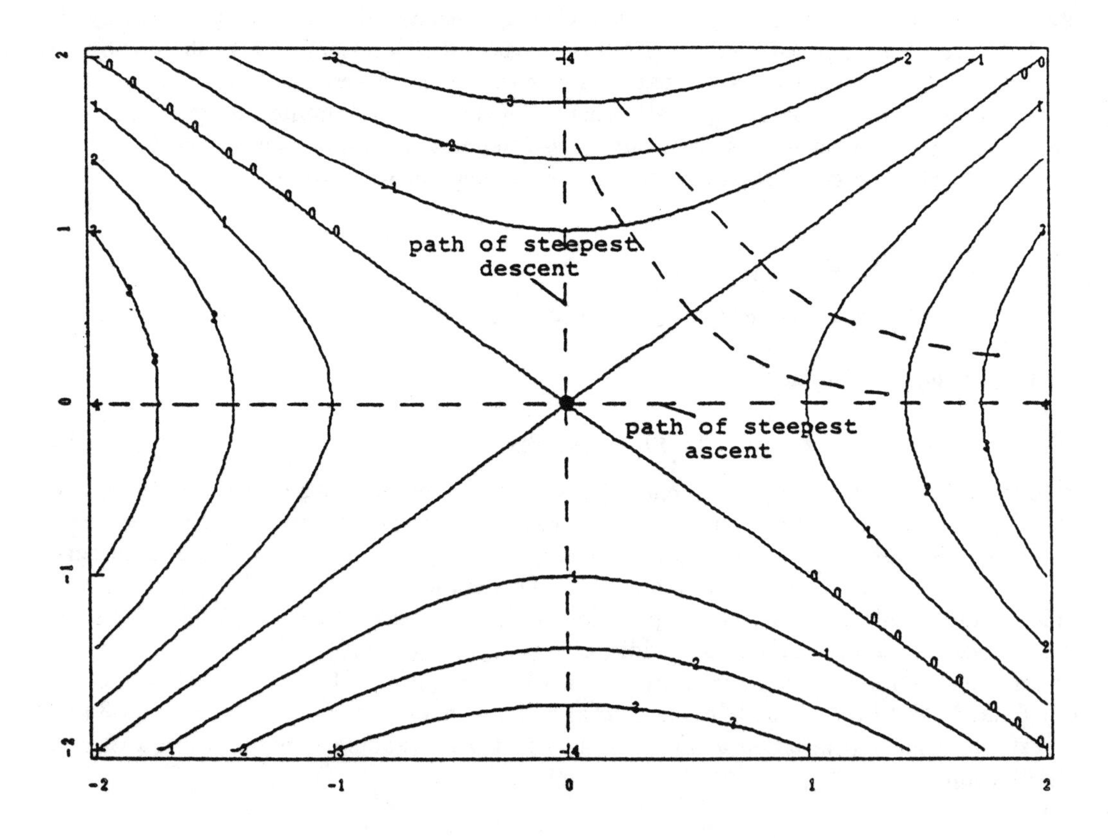


**Exhibit 3.1**

The path of steepest descent for the function

$w(z) = z^2 = u(x,y) + iv(x,y)$

$u(x,y) = x^2 - y^2;\ v(x,y) = 2xy$

$\gamma(x,y) = 0 - u(x,y) = y^2 - x^2$

Saddlepoint : $(x_0, y_0) = (0,0)$

———— : $u(x,y) =$ constant (level curves)

- - - - - : $v(x,y) =$ constant (orthogonal trajectories)

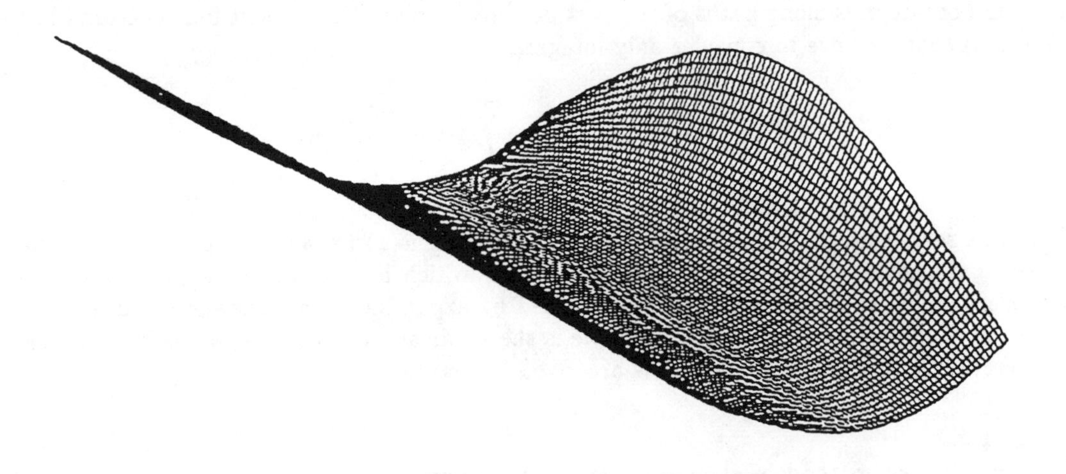

**Exhibit 3.2a**
Surface $u = u(x, y)$ for the function
$w(z) = u(x, y) + iv(x, y) = -\beta \log(1 - \frac{z}{\theta}) - z \cdot t$,
$t = 2$, $\theta = 0.5$, $\beta = 0.5$.

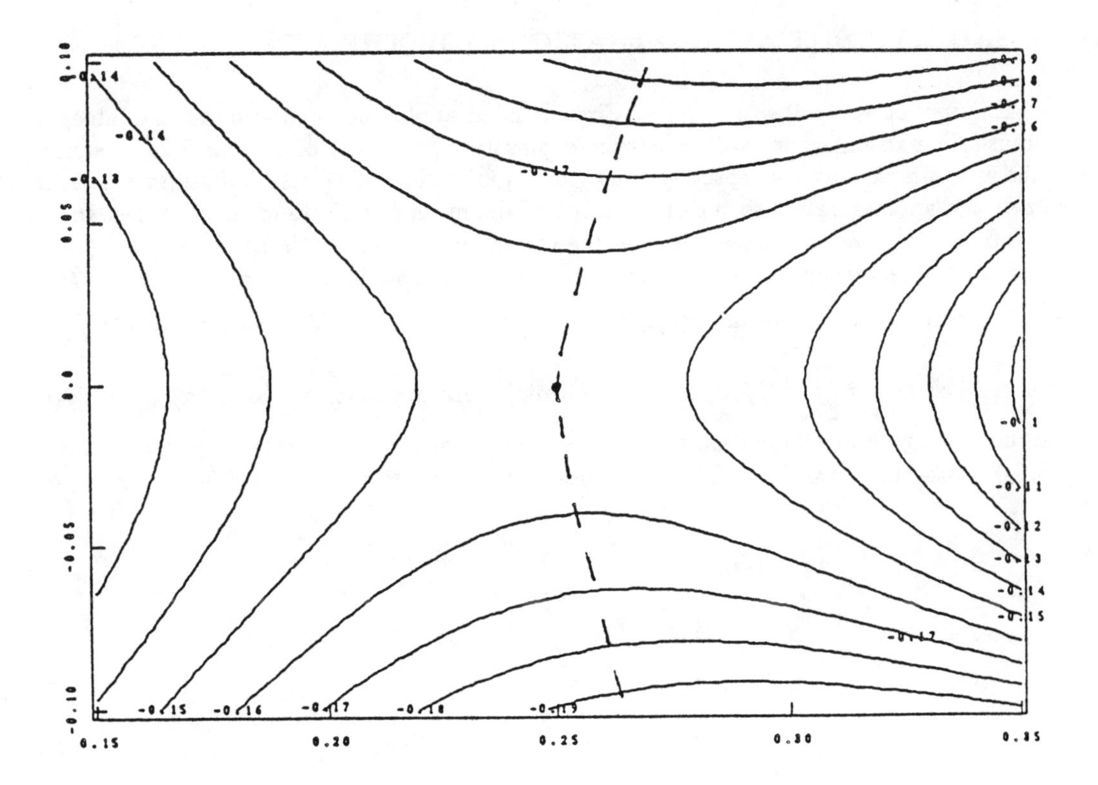


**Exhibit 3.2b**
Level curves and paths of steepest descent from the saddlepoint $(x_0, y_0) = (0.25, 0)$ for the function of Exhibit 3.2a.

To summarize: if it is possible to deform the path of integration and express the integral as a sum of integrals along paths of steepest descent from saddlepoints, it follows from (3.1) and (3.2) that we have to consider only integrals of the form

$$\int_{\mathcal{P}} e^{vw(z)}\xi(z)dz = e^{vw(z_0)} \int_0^{\infty} e^{-v\cdot\gamma}\xi(z)\frac{dz}{d\gamma}d\gamma. \tag{3.4}$$

It can be seen from (3.4) that instead of approximating $w(z)$ *in the exponential* (where the error would be blown up), we approximate $dz/d\gamma$ which has been moved down from the exponent. This approximation can be obtained by expanding this expression into a series near the saddlepoint $z_0$. A typical example is the application of this technique in statistics (see section 3.3). Further applications are given in chapter 7.

*Remark 3.1*
Historically, the method of steepest descent can be traced back to Riemann (1892) who found an asymptotic approximation to the hypergeometric function, that is a multiple of the integral (3.1) with $w(z) = \log[z(1-z)(1-sz)^{-1}]$ and $\xi(z) = z^a(1-z)^b(1-sz)^c$, where $a, b, c, s$ are real parameters and $\mathcal{P}$ is any curve which joins 0 and 1. Debye (1909) generalized the work of Riemann and obtained a complete asymptotic expansion for integrals of the form (3.1) by means of the idea presented in this section.

## 3.3. SADDLEPOINT APPROXIMATIONS FOR THE MEAN

This section serves two purposes. First, it is an application of the method of steepest descent. In particular, we will construct explicitly (i) and (ii) of section 3.2. Secondly, it shows a simple but conceptually important application of this technique in statistics, namely the approximation to the distribution of the mean of $n$ iid random variables. In our exposition in this section we will follow the outline of Daniels' (1954) fundamental paper. Given $n$ iid observations $x_1, \cdots, x_n$ with common known distribution $F(x)$ and density $f(x)$, we want to approximate the density $f_n(t)$ of the arithmetic mean $T_n(x_1, \cdots, x_n) = n^{-1}\sum_{i=1}^{n} x_i$. Denote by $M(\alpha) = \int_{-\infty}^{+\infty} e^{\alpha x} f(x)dx$ the moment generating function, by $K(\alpha) = \log M(\alpha)$ the cumulant generating function and suppose they exist for real values of $\alpha$ in some interval $(c_1, c_2)$ containing the origin. Then, by Fourier inversion the density $f_n(t)$ can be written as

$$\begin{aligned} f_n(t) &= (n/2\pi)\int_{-\infty}^{+\infty} M^n(ir)e^{-inrt}dr \\ &= (n/2\pi i)\int_{\Im} M^n(z)e^{-ntz}dz, \end{aligned} \tag{3.5}$$

where $\Im$ is the imaginary axis in the complex plane. Since the contributions of the integral over the paths $\mathcal{P}'$ and $\mathcal{P}''$ go to 0 as $\alpha \to \infty$ (see Exhibit 3.3), one can alternatively integrate over any straight line parallel to the imaginary axis. Therefore, (3.5) can be rewritten as

$$f_n(t) = (n/2\pi i) \int_{\tau - i\infty}^{\tau + i\infty} M^n(z) e^{-nzt} dz$$

$$= (n/2\pi i) \int_{\tau - i\infty}^{\tau + i\infty} \exp\{n[K(z) - zt]\} dz, \tag{3.6}$$

for any real number $\tau$ in the interval $(c_1, c_2)$.

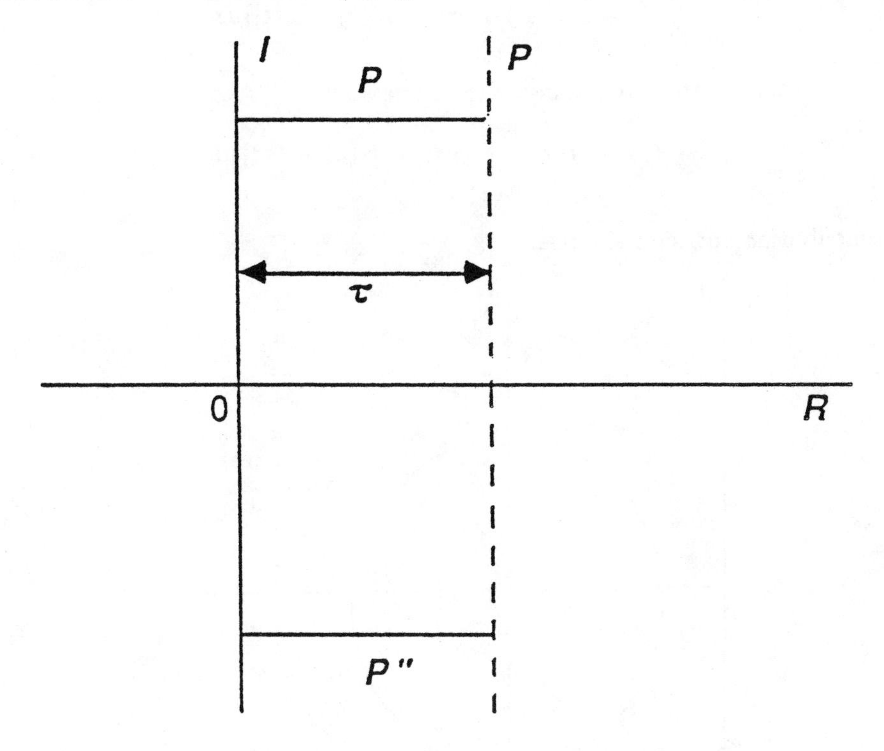


**Exhibit 3.3**

Shift of path of integration from $\Im$ to $\mathcal{P}$

At this point the integral in (3.6) is of the form (3.1) where $v = n$, $w(z) = K(z) - z \cdot t$ with $t$ fixed, $\xi(z) = n/2\pi i$, and the path $\mathcal{P}$ is a straight line parallel to the imaginary axis going through the real point $\tau$. Let us look at conditions (i) and (ii) of section 3.2. From (i) we see that the new path will have to go through a zero of $w'(z)$, that is

$$w'(z) = K'(z) - t = 0.$$

Thus, the new path will go through the saddlepoint $z_0$ defined as a solution to the equation

$$K'(z_0) = t.$$

Daniels (1954) shows in Theorems 6.1 and 6.2 under general conditions that the saddlepoint $z_0$ is ***unique and real*** on $(c_1, c_2)$, and that $K''(z_0) > 0$. Thus, from now on $z_0 = \alpha_0 \in \mathbf{R}$.

Condition (ii) requires that $\Im w(z) \equiv$ constant on the new path, that is $\Im w(z) \equiv \Im w(\alpha_0) = 0$ since $w(\alpha_0)$ is real. This allows us to deform $\mathcal{P}$ as shown in Exhibit 3.4.

First, choose $\tau = \alpha_0$ and move the integration path on the straight line parallel to the imaginary axis which goes through the real point $\alpha_0$. Secondly, construct a small circle of radius $\epsilon$ around the saddlepoint $\alpha_0$ and follow the path of steepest descent from the saddlepoint inside this circle $(\mathcal{P}_0)$. On this path $\Im w(z) = 0$ by condition (ii). Then, continue on the curves orthogonal to $(\mathcal{P}_0)$ at $z_1$ and $z_2$. Since $(\mathcal{P}_0)$ is a path of steepest descent, $(\mathcal{P}_1)$ and $(\mathcal{P}_2)$ are level curves defined by $\Re w(z) = \text{constant}$. From $z_3$ (respectively $z_4$) continue on the straight line $(\mathcal{P}_3, \mathcal{P}_4)$. Therefore, the original integral (3.5) can be rewritten as

$$f_n(t) = I_0 + I_1, \tag{3.7}$$

where

$$I_0 = (n/2\pi i) \int_{\mathcal{P}_0} \exp\{n[K(z) - z \cdot t]\} dz \tag{3.8}$$

is the contribution to the integral inside the circle and

$$I_1 = (n/2\pi i) \int_{\tilde{\mathcal{P}} \backslash \mathcal{P}_0} \exp\{n[K(z) - z \cdot t]\} dz \tag{3.9}$$

is the contribution outside the circle.

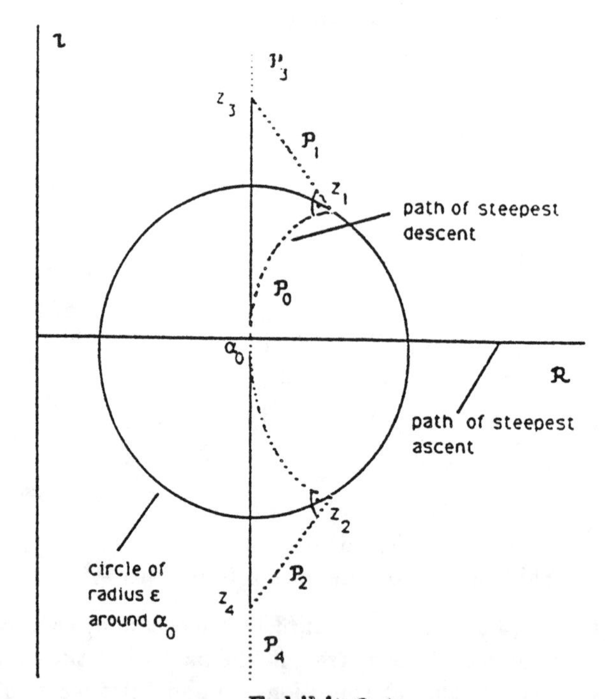


**Exhibit 3.4**
Deformation of the path of integration for the computation of the integral in (3.5).

Old path : imaginary axis

New path : $\tilde{\mathcal{P}} = \mathcal{P}_4 \cup \mathcal{P}_2 \cup \mathcal{P}_0 \cup \mathcal{P}_1 \cup \mathcal{P}_3 (- - - - -)$

$\mathcal{P}_0 : \Im(z) \equiv \Im w(\alpha_0) = 0$; this is a path of steepest descent which crosses $\Re$ orthogonally at $\alpha_0$.

$\Re : \Im w(z) \equiv 0$; this is a path of steepest ascent $(K''(\alpha_0) > 0)$

$\mathcal{P}_1 : \Re w(z) \equiv \text{ constant } = \Re w(z_3)$

$\mathcal{P}_2 : \Re w(z) \equiv \text{ constant } = \Re w(z_4)$

We first look at $I_1$. On the straight line $z = \alpha_0 + iy$ we have:

$$e^{w(z)} = M(z)e^{-zt} = \int \exp\{(\alpha_0 + iy)x\}dF(x) \cdot e^{-zt}$$

$$= \left\{\int e^{iyx}e^{\alpha_0 x}dF(x)/M(\alpha_0)\right\}M(\alpha_0)e^{-zt}$$

$$= \phi(y)M(\alpha_0)\exp\{-(\alpha_0 + iy)t\}, \tag{3.10}$$

where $\phi(y)$ is the characteristic function of a random variable with density $e^{\alpha_0 x}f(x)/M(\alpha_0)$. Therefore,

$$\left|e^{w(z)}\right| = \exp\{\Re w(z)\} \leq \rho \cdot e^{w(\alpha_0)},$$

with $\rho < 1$, and the contribution to the integral outside the circle on $\mathcal{P}_3$, and $\mathcal{P}_4$ is of order $0(\rho^n)$ and can be ignored. On $\mathcal{P}_1$ and $\mathcal{P}_2$, $\Re w(z)$ is constant and $e^{w(z)}$ can be bounded as above. Hence the contribution on $\mathcal{P}_1$ and $\mathcal{P}_2$ can be ignored.

Let us now look at $I_o$. By definition, $w(z)$ is real on $\mathcal{P}_o$. Define $\gamma(x,y)$ as in (3.3)

$$\gamma = w(\alpha_0) - w(z) = K(\alpha_0) - \alpha_0 \cdot t - [K(z) - zt]$$

and expand the right hand side in a series around $\alpha_0$

$$\gamma = -(z-\alpha_0)[K'(\alpha_0) - t] - (z-\alpha_0)^2K''(\alpha_0)/2$$

$$-(z-\alpha_0)^3K'''(\alpha_0)/6 - (z-\alpha_0)^4K^{(iv)}(\alpha_0)/24 - \cdots \tag{3.11}$$

Since $\gamma$ is real and steadily increasing from the saddlepoint, we rewrite it as $\gamma = \delta^2/2$. With the change of variable

$$\zeta = (z-\alpha_0)[K''(\alpha_0)]^{1/2}$$

and

$$\lambda_3(\alpha_0) = K'''(\alpha_0)/[K''(\alpha_0)]^{3/2}$$
$$\lambda_4(\alpha_0) = K^{(iv)}(\alpha_0)/[K''(\alpha_0)]^2,$$

and recalling that $K'(\alpha_0) = t$ we can rewrite (3.11) as

$$-\delta^2/2 = \zeta^2/2 + \lambda_3(\alpha_0)\zeta^3/6 + \lambda_4(\alpha_0)\zeta^4/24 + \cdots \tag{3.12}$$

The series (3.12) can be inverted in the neighborhood of $\zeta = 0$ ($z = \alpha_0$) that is $\zeta$ can be expressed as the following series of $\delta$

$$\zeta = i\delta + \lambda_3(\alpha_0)\delta^2/6 + \{\lambda_4(\alpha_0)/24 - (5/72)\lambda_3^2\}i\delta^3 + \cdots \tag{3.13}$$

At this point we are ready to rewrite $I_0$ according to (3.4). We obtain:

$$I_0 = (n/2\pi i)\int_{\mathcal{P}_0} \exp\{n[K(z) - zt]\}dz$$

$$= (n/2\pi i)\exp\{n[K(\alpha_o) - \alpha_o t]\}\int_{\mathcal{P}_0} e^{-n\gamma}dz$$

$$= (n/2\pi i)\frac{\exp\{n[K(\alpha_o) - \alpha_o t]\}}{[K''(\alpha_o)]^{1/2}}\int_{-A}^{B} e^{-n\delta^2/2}\frac{d\zeta}{d\delta}d\delta,$$

and from (3.13)

$$I_0 = (n/2\pi i)\frac{\exp\{n[K(\alpha_0) - \alpha_0 t]\}}{[K''(\alpha_0)]^{1/2}} \times$$

$$\int_{-A}^{B} e^{-n\delta^2/2}\Big\{i + \lambda_3(\alpha_0)\delta/3 + i\Big[\lambda_4(\alpha_0)/8 - \frac{5}{24}\lambda_3^2(\alpha_0)\Big]\delta^2 + \cdots\Big\}d\delta, \qquad (3.14)$$

where $A$ and $B$ are two positive numbers which correspond to the values of $\delta$ at $z_2$ and $z_1$. By applying Watson's Lemma (see below) to (3.14), one finally obtains the following asymptotic expansion

$$f_n(t) = \left[\frac{n}{2\pi K''(\alpha_0)}\right]^{1/2} \exp\{n[K(\alpha_0 t) - \alpha_0 t]\}$$

$$\times\left\{1 + \frac{1}{n}\left[\frac{1}{8}\lambda_4(\alpha_0) - \frac{5}{24}\lambda_3^2(\alpha_0)\right] + \cdots\right\} \qquad (3.15)$$

where $\alpha_0$ is determined by the *saddlepoint equation*

$$K'(\alpha_0) = t, \qquad (3.16)$$

and

$$\lambda_3(\alpha_0) = K'''(\alpha_0)/[K''(\alpha_0)]^{3/2} \qquad (3.17)$$

$$\lambda_4(\alpha_0) = K^{(iv)}(\alpha_0)/[K''(\alpha_0)]^2 \qquad (3.18)$$

are standardized measures of skewness and kurtosis respectively. The leading term of the expansion (3.15)

$$g_n(t) = \left[\frac{n}{2\pi K''(\alpha_0)}\right]^{1/2} \exp\{n[K(\alpha_0) - \alpha_0 t]\} \qquad (3.19)$$

is called the *saddlepoint approximation.*

For the sake of completeness we give here a modification of Watson's lemma due to Jeffreys and Jeffreys (1950) which is used in the final step in the derivation of (3.15).

**Lemma** (Watson, 1948; Jeffreys and Jeffreys, 1950; Daniels, 1954).

If $\psi(\zeta)$ is analytic in a neighborhood of $\zeta = 0$ and bounded for real $\zeta = \delta$ in an interval $-A \le \delta \le B$ with $A > 0$ and $B > 0$, then

$$(n/2\pi)^{1/2}\int_{-A}^{B} e^{-n\delta^2/2}\psi(\delta)d\delta \sim \psi(0) + \frac{1}{2n}\psi''(0) + \cdots + \frac{1}{(2n)^r}\frac{\psi^{(2r)}(0)}{r!} + \cdots$$

is an asymptotic expansion in powers of $n^{-1}$.

In the following remarks we discuss some aspects of the saddlepoint approximation in some detail.

*Remark 3.2:* Error of the approximation.
From $f_n(t) = g_n(t)[1 + 0(1/n)]$ one can see that $g_n(t) \geq 0$ and that the relative error is of order $n^{-1}$. This is the most important property of this approximation and a major advantage with respect to Edgeworth expansions. Moreover, Daniels (1954), p. 640 ff. showed that for a wide class of underlying densities, the coefficient of the term of order $n^{-1}$ doesn't depend on $t$. Thus, in such cases the relative error is of order $n^{-1}$ *uniformly*. cf. Jensen (1988).

Since $g_n(t)$ doesn't necessarily integrate to 1, one can renormalize the approximation by dividing by $C_n = \int g_n(t)dt$. This operation comes out naturally by using an alternative derivation of the saddlepoint approximation proposed by Hampel (1973) which is based on the expansion of $f_n'(t)/f_n(t)$ rather than $f_n(t)$; see sections 4.2 and 5.2. By renormalization one actually improves the order of the approximation by getting a relative error of order $0(n^{-3/2})$ for values $t$ in the range $t - \mu = 0(n^{-1/2})$, where $\mu = \int x dF(x)$. To see this write

$$f_n(t) = g_n(t)[1 + b(t)/n + 0(n^{-2})]$$

and

$$\begin{aligned} g_n(t) &= f_n(t)[1 - b(t)/n + 0(n^{-2})] \\ &= f_n(t)[1 - b(\mu)/n - (t-\mu)b'(\mu)/n + 0((t-\mu)^2/n) \\ &\quad + 0(n^{-2})] \\ &= f_n(t)[1 - b(\mu)/n - (t-\mu)b'(\mu)/n + 0(n^{-2})]. \end{aligned} \tag{3.20}$$

Therefore

$$C_n = \int g_n(t)dt = 1 - b(\mu)/n + 0(n^{-2}) \tag{3.21}$$

and from (3.20) and (3.21) by using $t - \mu = 0(n^{-1/2})$

$$g_n(t)/C_n = f_n(t)[1 + 0(n^{-3/2})].$$

If in addition the relative error is uniform of order $n^{-1}$, that is $b(t)$ does not depend on $t$, the relative error after renormalization is of order $0(n^{-2})$.

*Remark 3.3:* Exact saddlepoint approximations.
It turns out that in some cases the saddlepoint approximation $g_n(t)$ is exact or exact up to normalization. Daniels (1954,1980) proved that there are only three underlying densities $f(x)$ for which this is the case, namely the normal, the gamma, and the inverse normal distribution. In the case of the normal the leading term is exact and the higher order terms are zero. In the other two cases the leading term is exact up to a constant and the higher order terms are different from zero but independent of $t$ and can therefore be included in the normalization constant. Moreover, Blaesild and Jensen (1985) showed that $f(x)$ has an exact saddlepoint approximation if and only if $f(x)$ is a reproductive exponential model.

*Remark 3.4:* Computational issue.
In order to compute the saddlepoint approximation $g_n(t)$, one has to solve the implicit saddlepoint equation (3.16) for each value $t$. Since $K'(\cdot)$ has an integral form, this can be computational intensive in multidimensional problems; see chapter 4. However, if the density has to be approximated on an interval $[t_1, t_2]$ one can find the saddlepoint $\alpha_0^{(1)}$ corresponding to $t_1$ and use this as a starting point for the next value $t$, and so on. Moreover, when $\alpha_0$ as a function of $t$ is monotone as in the case of the mean and the density $f_n(t)$ does not have

to be approximated at equally spaced points, one can just take a number of values $\alpha_0$ and compute the corresponding values $t = K'(\alpha_0)$. cf. section 7.1.

*Remark 3.5:* Lattice underlying distribution.
A saddlepoint approximation for the mean can be derived when the underlying distribution is lattice; see Daniels (1983, 1987), Gamkrelidzt (1980).

The numerical examples in section 3.5 show the great accuracy of the saddlepoint approximation for the mean. The same pattern can be found for more important and complex situations; cf. chapters 4, 6 and 7.

## 3.4. RELATIONSHIP WITH THE METHOD OF CONJUGATE DISTRIBUTIONS

Up to this point, the approximation to the density of the mean $f_n(t)$ has been derived using the method of steepest descent and the saddlepoint approximation. In this section we develop the approximation using the idea of conjugate densities and the normal approximation. Although both approaches lead to the same results, we can gain new insight into the approximation through the conjugate density.

Probably the simplest and most common approximation to the density of the mean is the normal approximation. This approximation works very well near the center of the distribution but breaks down in the tail. The idea here is to re-center our density at the point of interest $t$ by means of a conjugate density and then to use a normal approximation in the re-centered problem. The approximation in the re-centered problem is then converted to an approximation for $f_n(t)$.

To be more specific, we introduce the conjugate density

$$h_t(x) = c(t)\exp\{\alpha(t)(x-t)\}f(x) \tag{3.22}$$

where $c(t)$ is chosen so that $h_t$ is a density and $\alpha(t)$ is chosen so that

$$\int (x-t)\exp\{\alpha(t)(x-t)\}f(x)dx = 0 \qquad i.e.\ E_{h_t}X = t. \tag{3.23}$$

The variance of $X$ under $h_t$ is denoted by $\sigma^2(t)$, i.e.

$$\sigma^2(t) = c(t)\int (x-t)^2\exp\{\alpha(t)(x-t)\}f(x)dx.$$

Conjugate or associated densities are well known in probability (Khinchin, 1949; Feller, 1971, p. 518) and arise naturally in information theory (Kullback 1960).

To illustrate the way in which the conjugate density operates, it is informative to look at some graphs. Exhibit 3.5 shows the situation when $f(x)$ is uniform on $[-1,1]$.
The plots give the conjugate densities centered at .3, .5, .7, .9. As we move to the right $\alpha(t)$ increases in order to put sufficient mass in the interval $[t,1]$. The conjugate is plotted only for the interval $[0,1]$. On $[-1,0]$ all four conjugates are relatively flat and close together.

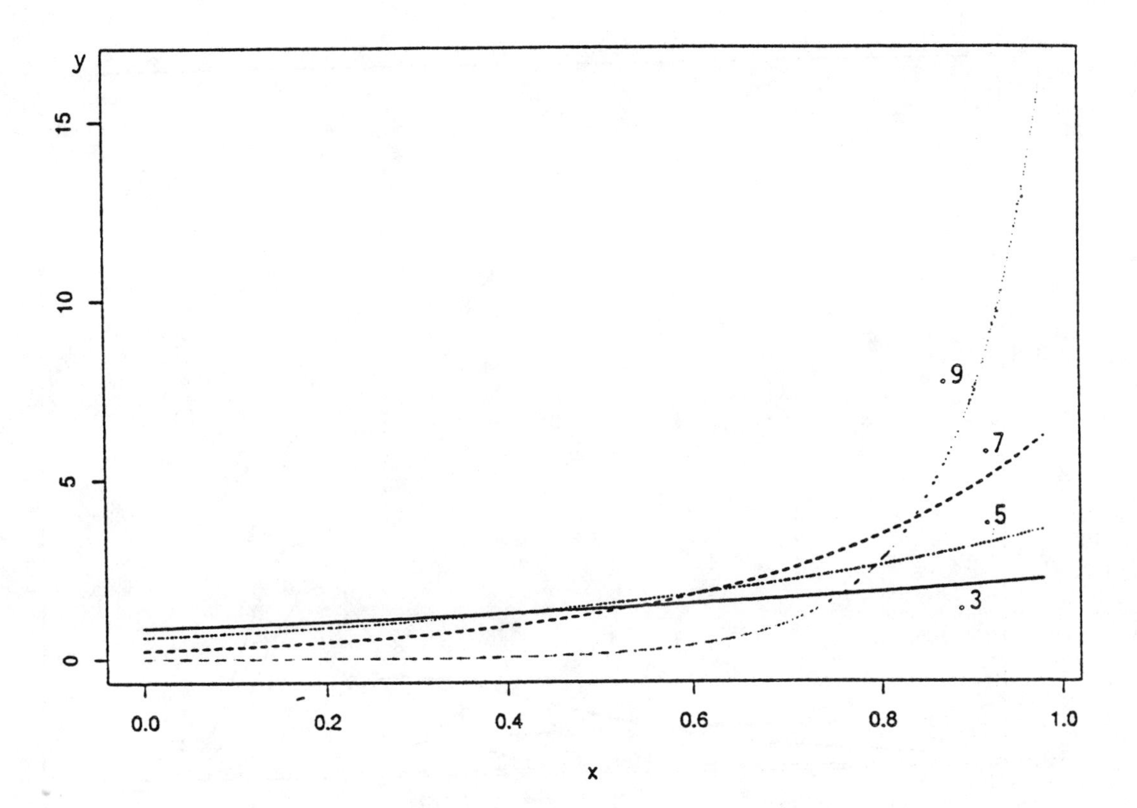


**Exhibit 3.5**
Uniform conjugate at $t = .3, .5, .7, .9$.

As a second example consider the extreme density with $f(x) = \exp\{x - \exp(x)\}$. Conjugate densities are plotted in Exhibit 3.6 for $t$ values $-7$, $-3$, 0, 0.5, 2 along with $f$. As can be seen the shape of the density is changed substantially as the values of $t$ are varied.

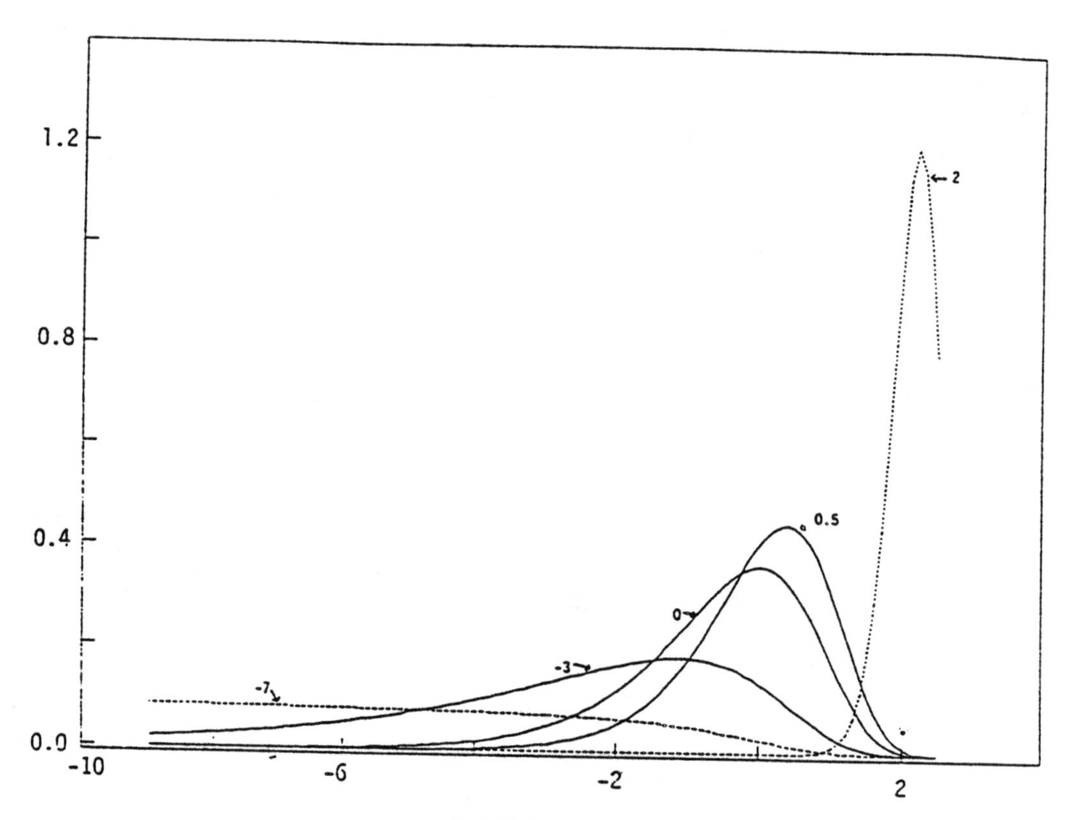


**Exhibit 3.6**
Extreme conjugate at $t = -7, -3, 0, 0.5, 2$.

Before proceeding with the development of the approximation, it is important to link the notation of the conjugate density with that of the cumulant generating functions that have been used up to now. Recall from section 3.3 that

$$K(\alpha) = \log \int e^{\alpha x} f(x)dx$$

and the saddlepoint (at a point $t$) is the solution of $K'(\alpha_0) = t$ (see (3.16))

$$\int x \exp(\alpha_0 x) f(x)dx \Big/ \int \exp(\alpha_0 x) f(x)dx = t$$

or

$$\int (x-t) \exp(\alpha_0 x) f(x)dx \Big/ \int \exp(\alpha_0 x) f(x)dx = 0.$$

Multiplying both numerator and denominator by $\exp(-\alpha_0 t)$ we have

$$\int (x-t) \exp\{\alpha_0(x-t)\} f(x)dx \Big/ \int \exp\{\alpha_0(x-t)\} f(x)dx = 0$$

or

$$\int (x-t) \exp\{\alpha_0(x-t)\} f(x)dx = 0.$$

By comparing this with (3.23), we see that $\alpha_0 = \alpha(t)$, i.e. $\alpha(t)$ is the saddlepoint at $t$. Moreover, $c^{-1}(t) = \int \exp\{\alpha(t)(x-t)\} f(x)dx = \exp\{K(\alpha(t)) - \alpha(t)t\}$, hence $-\log c(t) =$

$K(\alpha(t)) - \alpha(t)t = K(\alpha_0) - \alpha_0 t$. Similarly

$$K''(\alpha_0) = K''(\alpha(t))$$
$$= \frac{\int \exp\{\alpha(t)x\}f(x)dx \int x^2 \exp\{\alpha(t)x\}f(x)dx - \left(\int x \exp\{\alpha(t)x\}f(x)dx\right)^2}{\left(\int \exp\{\alpha(t)x\}f(x)dx\right)^2}$$

Multiplying numerator and denominator by $(e^{-\alpha(t)t})^2$ we have

$$K''(\alpha(t)) = E_{h_t}X^2 - (E_{h_t}X)^2 = \sigma^2(t).$$

To summarize, we have

$$\alpha(t) = \alpha_0, \quad -\log c(t) = K(\alpha_0) - \alpha_0 t, \quad \sigma^2(t) = K''(\alpha_0). \tag{3.24}$$

The next step is to consider the density of the mean under $h_t$, say $h_{t,n}$ and relate it to $f_n$, the density of the mean under $f$. We can write $f_n$ as follows:

$$f_n(t) = n \int \cdots \int f\left(nt - \sum_1^{n-1} x_i\right) \prod_1^{n-1} f(x_i)dx_1 \cdots dx_{n-1}$$
$$= c^{-n}(t)n \int \cdots \int c(t)\exp\left\{\alpha(t)\left(nt - \sum_1^{n-1} x_i - t\right)\right\} f\left(nt - \sum_1^{n-1} x_i\right)$$
$$\prod_1^{n-1} c(t)\exp\{\alpha(t)(x_i - t)\}f(x_i)dx_1 \cdots dx_{n-1}$$
$$= c^{-n}(t)n \int \cdots \int h_t\left(nt - \sum_1^{n-1} x_i\right) \prod_1^{n-1} h_t(x_i)dx_1 \cdots dx_{n-1}$$
$$= c^{-n}(t)h_{t,n}(t).$$

Hence we conclude

$$f_n(t) = c^{-n}(t)h_{t,n}(t) \tag{3.25}$$

This centering equation provides the link for relating the approximation to $h_{t,n}(t)$ to the desired approximation of $f_n(t)$.

The final step is to approximate $h_{t,n}(t)$. Now $h_{t,n}(t)$ is the density of $\bar{X}$ under $h_t$ where the $X_i$'s have mean $t$ and variance $\sigma^2(t)$. Expression (2.10) gives the Edgeworth expansion at the origin for a standardized variable with

$$\lambda_3 \equiv \lambda_3(t) = E_{h_t}(X-t)^3/\sigma^3(t)$$
$$\lambda_4 \equiv \lambda_4(t) = E_{h_t}(X-t)^4/\sigma^4(t).$$

The approximation to the density of $\sqrt{n}(\bar{X} - t)/\sigma(t)$ at 0 is

$$\frac{1}{\sqrt{2\pi}}\left[1 + \frac{1}{n}\left(\frac{\lambda_4(t)}{8} - \frac{5\lambda_3^2(t)}{24}\right) + 0\left(\frac{1}{n^2}\right)\right].$$

From this it follows that

$$h_{t,n}(t) = \sqrt{\frac{n}{2\pi}}\ \frac{1}{\sigma(t)}\left(1+0\left(\frac{1}{n}\right)\right)$$

and

$$f_n(t) = \sqrt{\frac{n}{2\pi}}\ \frac{c^{-n}(t)}{\sigma(t)}\left(1+0\left(\frac{1}{n}\right)\right).$$

This leads to the small sample (or saddlepoint) approximation for the mean

$$g_n(t) = \sqrt{\frac{n}{2\pi}}\ c^{-n}(t)/\sigma(t). \tag{3.26}$$

Using (3.24), it can be seen that this is exactly (3.19).

It should be noted that (3.26) is obtained by shifting the underlying density to the point of interest, using a normal approximation at the mean, and then using the centering lemma. The process can be likened to using a low order Taylor's expansion at many points rather than one high order expansion at a single point as in an Edgeworth expansion. The approximation (3.26) can be thought of as a local normal approximation.

It is worth asking whether the form of the conjugate density is important for the argument above. If we look at the argument leading to the centering lemma, the exponential in the conjugate enabled us to go from $f$ to $h_t$. It is hard to see how to obtain the necessary link between $f_n$ and $h_{t,n}$ with any other form. From another point of view, start with the density $f$ and ask for the density closest to $f$ in Kullback-Liebler distance which is constrained to have mean $t$. Kullback (1960) shows that $h_t(x)$ is this density. By using the conjugate density, we have embedded our problem within an exponential family. This perspective becomes very useful in applying our techniques in multiparameter problems as we show in chapter 6.

### 3.5. EXAMPLES

We now consider numerical results from using approximation (3.26) with several underlying densities. In each case, the $\alpha(t)$ has been evaluated at a grid suitable for the underlying density. To solve this non-linear equation, we have used a secant style root finder, such as C05AJ5 in the NAG library. Given $\alpha(t)$, $c(t)$ and $\sigma(t)$ can be determined using numerical integration.

As a first example, let $f(x)$ be uniform on $[-1,1]$ and consider the density of the mean for $n = 5$. The exact density is given by

$$f_n(t) = \frac{n^n}{2^n(n-1)!}\sum_{i=0}^{n}(-1)^i\binom{n}{i}\left\langle 1-t-\frac{2i}{n}\right\rangle^{n-1} \qquad |t| \leq 1$$

where $< z >= z$ for $z \geq 0$ and $= 0$ for $z < 0$, cf. section 2.7. Exhibit 3.7 gives the exact and approximate density for some selected points. The error is measured by percent relative error = 100 (approximate-exact)/exact.

As can be seen the results are very accurate even for the extreme tail. They are clearly superior to the approximations obtained by Edgeworth expansions (see section 2.7). The maximum percent relative error is 1.65% for $t = .95$ and is actually smaller for values beyond .95. Such accuracy would certainly be more than adequate for all applications. The property that the percent relative error stays bounded even in the extreme tail seems to be quite general with these approximations.

| t | Exact | Approximate | % relative error |
|---|---|---|---|
| 0.00 | 1.4974e0 | 1.4945e0 | -0.19 |
| 0.05 | 1.4731e0 | 1.4700e0 | -0.21 |
| 0.10 | 1.4022e0 | 1.3988e0 | -0.24 |
| 0.15 | 1.2900e0 | 1.2872e0 | -0.22 |
| 0.20 | 1.1458e0 | 1.1446e0 | 0.10 |
| 0.25 | 9.8216e-1 | 9.8277e-1 | 0.06 |
| 0.30 | 8.1217e-1 | 8.1361e-1 | 0.18 |
| 0.35 | 6.4687e-1 | 6.4836e-1 | 0.23 |
| 0.40 | 4.9479e-1 | 4.9622e-1 | 0.29 |
| 0.45 | 3.6204e-1 | 3.6369e-1 | 0.46 |
| 0.50 | 2.5228e-1 | 2.5428e-1 | 0.79 |
| 0.55 | 1.6673e-1 | 1.6872e-1 | 1.19 |
| 0.60 | 1.0417e-1 | 1.0548e-1 | 1.26 |
| 0.65 | 6.1061e-2 | 6.1494e-2 | 0.71 |
| 0.70 | 3.2959e-2 | 3.2902e-2 | -0.17 |
| 0.75 | 1.5895e-2 | 1.5736e-2 | -1.00 |
| 0.80 | 6.5104e-3 | 6.4131e-3 | -1.49 |
| 0.85 | 2.0599e-3 | 2.0262e-3 | -1.64 |
| 0.90 | 4.0690e-4 | 4.0085e-4 | -1.49 |
| 0.95 | 2.5431e-5 | 2.5011e-5 | -1.65 |
| 0.96 | 1.0417e-5 | 1.0245e-5 | -1.65 |
| 0.97 | 3.2959e-6 | 3.2422e-6 | -1.63 |
| 0.98 | 6.5105e-7 | 6.4780e-7 | -0.50 |

**Exhibit 3.7**
Exact and approximate density for mean with uniform on $[-1,1]$ and $n=5$.

In many situations, it is not the density itself which is of interest but rather the tail area. In order to compute the tail area, it is possible to integrate the approximate density over a grid of points. However it is much easier and equally accurate to use a tail area approximation developed by Lugannani and Rice (1980) and discussed by Daniels (1987) and Tingley (1987). This approximation is discussed in some detail in section 6.1 but for completeness we give it here.

$$P[\bar{X} \geq t] \approx 1 - \Phi((2n \log c(t))^{1/2}) + c^{-n}(t)\left[\frac{1}{(\sigma(t)(\alpha(t)))} - \frac{1}{(2\log c(t))^{1/2}}\right]/(n2\pi)^{1/2} \qquad (3.27)$$

The approximation is very useful since to get the tail beyond $t$, we need only evaluate $\alpha(t)$, $\sigma(t)$ and $c(t)$. The numerical evidence is that the approximation gives an accuracy comparable to that obtained by numerical integration (cf. Daniels, 1987).

For the uniform case, Exhibit 3.8 gives the exact and approximate tail areas. Because of symmetry, only upper tail areas are given.

| t | Exact | Approximate | % relative error |
|---|---|---|---|
| 0.05 | 4.2554e-1 | 4.2547e-1 | -0.01 |
| 0.10 | 3.5347e-1 | 3.5338e-1 | -0.02 |
| 0.15 | 2.8601e-1 | 2.8593e-1 | -0.03 |
| 0.20 | 2.2500e-1 | 2.2493e-1 | -0.03 |
| 0.25 | 1.7175e-1 | 1.7165e-1 | -0.06 |
| 0.30 | 1.2689e-1 | 1.2675e-1 | -0.11 |
| 0.35 | 9.0451e-2 | 9.0291e-2 | -0.18 |
| 0.40 | 6.1979e-2 | 6.1837e-2 | -0.23 |
| 0.45 | 4.0648e-2 | 4.0537e-2 | -0.27 |
| 0.50 | 2.5391e-2 | 2.5298e-2 | -0.36 |
| 0.55 | 1.5016e-2 | 1.4922e-2 | -0.63 |
| 0.60 | 8.3333e-3 | 8.2371e-3 | -1.15 |
| 0.65 | 4.2742e-3 | 4.1961e-3 | -1.83 |
| 0.70 | 1.9775e-3 | 1.9307e-3 | -1.37 |
| 0.75 | 7.9473e-4 | 7.7478e-4 | -2.51 |
| 0.80 | 2.6042e-4 | 2.5486e-4 | -2.13 |
| 0.85 | 6.1798e-5 | 6.0932e-5 | -1.40 |
| 0.90 | 8.1380e-6 | 8.0784e-6 | -0.73 |

**Exhibit 3.8**
Exact and approximate tail areas for mean with uniform on $[-1,1]$ and $n = 5$.

As might be expected from the results with the density, we obtain the same order of accuracy for the tail areas. Again the relative error is much smaller than that of the Edgeworth approximation where it can reach 20%; cf section 2.7. These results indicate that the saddlepoint approximation gives very accurate results for small values of $n$.

To give some further numerical results, we now turn to the extreme case of $n = 1$. Since our expansion is asymptotic, there is no a priori reason to believe the approximation should work well in this case. For $n = 1$, the approximation (3.26) simply approximates the underlying density $f$. This point is discussed further in section 7.2. The convenient feature with $n = 1$ is that we can easily compute the exact results as a comparison. As a first example, we consider the case of the extreme density, $f(x) = \exp(x - \exp(x))$. Since $f$ is asymmetric, we consider behavior in the upper and lower tails. The results are given in Exhibit 3.9.

| t | Exact | Approximate | % relative error |
|---|---|---|---|
| -9.0 | 1.2339e-4 | 1.5693e-4 | 27.18 |
| -8.0 | 3.3535e-4 | 4.4526e-4 | 32.78 |
| -7.0 | 9.1105e-4 | 1.2270e-3 | 34.68 |
| -6.0 | 2.4726e-3 | 3.2752e-3 | 32.46 |
| -5.0 | 6.6927e-3 | 8.4597e-3 | 26.40 |
| -4.0 | 1.7983e-2 | 2.1138e-2 | 17.54 |
| -3.0 | 4.7369e-2 | 5.1054e-2 | 7.78 |
| -2.5 | 7.5616e-2 | 7.8207e-2 | 3.43 |
| -2.0 | 1.1821e-1 | 1.1818e-1 | -0.02 |
| -1.5 | 1.7851e-1 | 1.7451e-1 | -2.24 |
| -1.0 | 2.5465e-1 | 2.4639e-1 | -3.24 |
| -0.5 | 3.3070e-1 | 3.1970e-1 | -3.33 |
| 0.0 | 3.6788e-1 | 3.6091e-1 | -1.89 |
| 0.5 | 3.1704e-1 | 3.3015e-1 | 4.13 |
| 1.0 | 1.7937e-1 | 1.6912e-1 | -5.71 |
| 1.5 | 5.0707e-2 | 4.0221e-2 | -20.68 |
| 2.0 | 4.5663e-3 | 6.6205e-3 | 44.99 |
| 2.5 | 6.2366e-5 | 2.8859e-5 | -53.73 |

**Exhibit 3.9**

Exact and approximate density for the extreme for $n = 1$.

Certainly the results are not as accurate as with $n = 5$. However the approximate density gives fairly reasonable results even as we go out into the tails. We can obtain another view by examining tail areas for the extreme with $n = 1$. The values for the negative $t$ are lower tail areas and are upper tail errors for $t \geq 0$ (Exhibit 3.10).

| t | Exact | Approximate | % relative error |
|---|---|---|---|
| -9.0 | 1.2338e-4 | 1.2382e-4 | -0.35 |
| -8.0 | 3.3540e-4 | 3.7148e-4 | -9.71 |
| -7.0 | 9.1147e-4 | 1.0723e-3 | -15.00 |
| -6.0 | 2.4757e-3 | 2.9806e-3 | -16.94 |
| -5.0 | 6.7153e-3 | 7.9918e-3 | -15.97 |
| -4.0 | 1.8149e-2 | 2.0710e-2 | -12.37 |
| -3.0 | 4.8568e-2 | 5.2058e-2 | -6.70 |
| -2.5 | 7.8806e-2 | 8.1733e-2 | -3.58 |
| -2.0 | 1.2658e-1 | 1.2751e-1 | -0.73 |
| -1.5 | 1.9999e-1 | 1.9662e-1 | 1.71 |
| -1.0 | 3.0780e-1 | 2.8111e-1 | 9.49 |
| 0.0 | 3.6788e-1 | 3.7455e-1 | -1.78 |
| 0.5 | 1.9230e-1 | 2.0463e-1 | -6.03 |
| 1.0 | 6.5988e-2 | 5.9525e-2 | 10.86 |
| 1.5 | 1.1314e-2 | 9.7135e-3 | 16.48 |
| 2.0 | 6.1798e-4 | 7.7569e-4 | -20.33 |
| 2.5 | 5.1193e-6 | 2.4462e-6 | 109.28 |

**Exhibit 3.10**

Exact and approximate tail area for extreme with $n = 1$.

The results here are remarkably good except in the extreme upper tail. To complete, the section we look at the results with $n = 1$ for three densities; the uniform on $[-1, 1]$ (Exhibit 3.11), a sum of exponentials where the number in the sum is Poisson with parameter 4 (Exhibit 3.12) and a density $f(x) = 1 + \cos 4\pi x$ on $[0, 1]$ (Exhibit 3.13). These were chosen to show the varying degrees of accuracy one can get.

| t | Exact | Approximate | % relative error |
|---|---|---|---|
| 0.05 | 0.475 | 0.4706 | -0.92 |
| 0.10 | 0.450 | 0.4414 | -1.91 |
| 0.15 | 0.425 | 0.4124 | -2.96 |
| 0.20 | 0.400 | 0.3838 | -4.05 |
| 0.25 | 0.375 | 0.3556 | -5.16 |
| 0.30 | 0.350 | 0.3281 | -6.27 |
| 0.35 | 0.325 | 0.3011 | -7.34 |
| 0.40 | 0.300 | 0.2750 | -8.34 |
| 0.45 | 0.275 | 0.2496 | -9.22 |
| 0.50 | 0.250 | 0.2252 | -9.91 |
| 0.55 | 0.225 | 0.2017 | -10.35 |
| 0.60 | 0.200 | 0.1791 | -10.44 |
| 0.65 | 0.175 | 0.1574 | -10.07 |
| 0.70 | 0.150 | 0.1363 | -9.11 |
| 0.75 | 0.125 | 0.1156 | -7.48 |
| 0.80 | 0.100 | 0.09476 | -5.24 |
| 0.85 | 0.075 | 0.07300 | -2.66 |
| 0.90 | 0.050 | 0.05002 | 0.05 |
| 0.95 | 0.025 | 0.02562 | 2.47 |
| 0.96 | 0.020 | 0.02061 | 3.03 |
| 0.97 | 0.015 | 0.01555 | 3.65 |
| 0.98 | 0.010 | 0.01054 | 5.40 |

**Exhibit 3.11**
Exact and approximate tail area for uniform with $n = 1$.

| t | Exact | Approximate | % relative error |
|---|---|---|---|
| 5 | 0.3070 | 0.3079 | -0.30 |
| 7 | 0.1425 | 0.1430 | -0.32 |
| 9 | 0.05950 | 0.05961 | -0.33 |
| 11 | 0.02277 | 0.02284 | -0.34 |
| 12 | 0.01373 | 0.01378 | -0.35 |
| 13 | 0.008151 | 0.008180 | -0.36 |
| 14 | 0.004773 | 0.004790 | -0.36 |
| 15 | 0.002759 | 0.002769 | -0.38 |
| 16 | 0.001576 | 0.001582 | -0.39 |
| 17 | 0.000890 | 0.000894 | -0.41 |

**Exhibit 3.12**
Exact and approximate tail area for Poisson sum of exponentials with $n = 1$.

| t | Exact | Approximate | % relative error |
|---|---|---|---|
| 0.55 | 0.4032 | 0.4443 | 10.19 |
| 0.60 | 0.3243 | 0.3895 | 20.10 |
| 0.65 | 0.2743 | 0.3365 | 22.66 |
| 0.70 | 0.2532 | 0.2860 | 12.96 |
| 0.75 | 0.2500 | 0.2390 | - 4.41 |
| 0.80 | 0.2468 | 0.1960 | -20.57 |
| 0.85 | 0.2257 | 0.1581 | -29.94 |
| 0.90 | 0.1757 | 0.1273 | -27.51 |
| 0.95 | 0.09677 | 0.1134 | 17.15 |

**Exhibit 3.13**
Exact and approximate tail area for $f(x) = 1 + \cos 4\pi x$
on $[0, 1]$ with $n = 1$.

From the displays, we can see that the approximation is very accurate for the Poisson sum of exponentials, quite accurate for the uniform and less accurate for the cosine density. Since the approximation is based on a local normal approximation, the quality of the approximation is determined by the degree to which $(\bar{X} - t)/\sigma(t)$ under $h_t$ is approximated by a normal. In some recent work, Field and Massam (1987) have developed a diagnostic function for the accuracy of the approximation.

We conclude from Exhibit 3.13 that even by taking multimodal $f$ and choosing $n = 1$, we cannot make the approximation breakdown. Our evidence is that we have an asymptotic approximation which gives reasonable results for $n = 1$.

# 4. GENERAL SADDLEPOINT APPROXIMATIONS

## 4.1. INTRODUCTION

In this chapter, we use the approximations obtained for the mean to get approximations for more complicated statistics. The first section considers one-dimensional M-estimates i.e. $T_n$ is the solution of $\sum_1^n \psi(x_i;t) = 0$. The estimate $T_n$ is written locally as a mean and the saddlepoint approximation for the mean is used. In section 4.3, we consider a slightly different approach in that the moment generating function is approximated and a saddlepoint approximation used. The technique is applied to approximating the density of L-estimates in the next section. At this point, we turn to the problem for multivariate M-estimates. Techniques are similar to those used for one-dimensional M-estimates. Finally we modify the results to handle the case of regression using M-estimates. Throughout the chapter, there are numerical results illustrating the accuracy of these approximations even for small sample sizes.

In the cases considered in this chapter, our interest is to be able to say something about the density of an estimate. Although asymptotic results are available in most cases, we usually do not know whether these asymptotic distributions are good approximations for small or moderate sample sizes. For instance there are several proposals for using "t-statistics" based on robust location/scale estimates as a means for computing confidence intervals. We then need to know whether this t-approximation works reasonably and if it does, what are appropriate degrees of freedom. Some results in this direction are given in section 4.5.b.

## 4.2. ONE-DIMENSIONAL M-ESTIMATORS

To begin, consider the problem of finding a saddlepoint approximation for the density of a one-dimensional M-estimate. As developed by Huber (1964, 1967) M-estimates are defined as the solution $T_n$ of

$$\sum_{i=1}^{n} \psi(x_i, t) = 0 \tag{4.1}$$

for observations $x_1, x_2, \cdots, x_n$. If the $x_i$'s are independent observations from a density $f(x, \theta)$, then by setting $\psi(x, \theta) = \frac{\partial}{\partial \theta} \log f(x, \theta)$, $T_n$ becomes the maximum likelihood estimate of $\theta$. In much of the work on robustness, M-estimates play a central role. However the derivation of the exact density of such an estimate is usually intractable mathematically and it becomes essential to have a good approximation in order to carry out inference.

Denote the density of $T_n$ when the $x_i$'s are independent observations from a density $f$ as $f_n(t)$. To approximate $f_n(t)$, we proceed by writing $T_n$ as a mean up to a certain order and then using the saddlepoint approximation to the mean as derived in section 3.2. The approach follows closely that developed in Field (1982) for multivariate M-estimates which in turn uses critically results on multivariate Edgeworth expansions in Bhattacharya and Ghosh (1978). Field and Hampel (1982) give an alternate derivation for univariate M-estimates based on the log-derivative density. In this section, the approximation is developed for a one-dimensional M-estimate. The development for multivariate M-estimates is presented in section 4.5 .

In the development of the approximation, the conjugate density (cf. (3.22)) will play a central role. For the case of an M-estimate the conjugate density for a fixed value of $t$ is defined as

$$h_t(x,\theta) = c(t)\exp\{\alpha(t)\psi(x,t)\}f(x,\theta) \tag{4.2}$$

By an appropriate choice of $\alpha(t)$, we will use the conjugate density to center $T_n$ at $t$ in the sense that $E_{h_t}(T_n) = t$ up to first order. This enables us to use a low order Edgeworth expansion to approximate $f_n$ at $t$. $\alpha(t)$ will be chosen so that

$$\int \psi(x,t)\exp\{\alpha(t)\psi(x,t)\}f(x,\theta)dy = 0. \tag{4.3}$$

The following assumptions on $\psi$ and $f(x,\theta)$ will be required in the development of the approximation. $D^v$ denotes the $v^{th}$ derivative of $\psi(x,\theta)$ with respect to $\theta$.

A4.1 The equation (4.1) has a unique solution $T_n$ and equation (4.2) has a unique solution $\alpha(t)$.

A4.2 There is an open subset $U$ of $R$ such that

(i) for each $\theta \in \Theta$, $F_\theta(U) = 1$

(ii) $D\psi(x,\theta)$, $D^2\psi(x,\theta)$, $D^3\psi(x,\theta)$ exist.

A4.3 For each compact $K \subset \Theta$

(i) $$\sup_{\theta_0 \in K} E_{\theta_0}|D^2\psi(X,\theta_0)|^4 < \infty$$

(ii) there is an $\epsilon > 0$ such that

$$\sup_{\theta_0 \in K} E_{\theta_0}(\max_{|\theta-\theta_0|\le\epsilon} |D^3\psi(X,\theta)|^3) < \infty$$

A4.4 For each $\theta_0 \in \Theta$, $E_{\theta_0}\psi(X,\theta_0) = 0$ and

$$A(\theta_0) = E_{\theta_0}[D\psi(X,\theta_0)] \neq 0.$$

A4.5 The functions $A(\theta)$ and $E_\theta\{[D^2\psi(X,\theta)]^2\}$ are continuous on $\Theta$.

The approach is to now fix a point $t_o$ at which to approximate $f_n$ and construct the conjugate density for $t_o$ as in (4.2) and (4.3). The next step is to approximate the density of $T_n$ under the conjugate density. As we will show, it will suffice to use a normal approximation at this point. To complete the process, we need a result linking the density of $T_n$ evaluated under the conjugate $h_{t_o}$ with that under the density $f$. The following centering result provides that link. In order to simplify the proof of the theorem, we add the following assumption

A4.6 $|D\psi(x,\theta)|$ is bounded by $k$.

A proof of the theorem which does not require the assumption is given in Field (1982, p. 673). In the following we suppress the dependence of the density $f$ on $\theta$.

**Theorem 4.1**

If assumptions A4.1 and A4.6 hold, then

$$f_n(t_o) = c^{-n}(t_o)h_{t_o,n}(t_o) \tag{4.4}$$

where $h_{t_o,n}(t)$ is the density of $T_n$ under the conjugate density $h_{t_o}(x.\theta)$ and

$$c^{-1}(t_0) = \int \exp\{\alpha(t_0)f(x,\theta)\}dx.$$

Proof: To illustrate the ideas, we first give a proof for the discrete case

$$f_n(t_0) = P_f[T_n = t_0] = \sum_{\{\mathbf{x} | \sum \psi(x_i, t_0) = 0\}} \prod_1^n f(x_i)$$

$$= \sum_{\{\mathbf{x} | \sum \psi(x_i, t_0) = 0\}} c^{-n}(t_0) \prod_1^n f(x_i) \exp\{\alpha(t_0)\psi(x_i, t_0)\} c(t_0)$$

$$= c^{-n}(t_0) P_{h_{t_0}}[T_n = t_0] = c^{-n}(t_0) h_{t_0,n}(t_0).$$

In the continuous case, let

$$A(\Delta t) = \{\mathbf{x} | \sum_{i=1}^n \psi(x_i, u) = 0, \ \ t_0 \leq u \leq t_0 + \Delta t\}.$$

Now

$$f_n(t_0) = \lim_{\Delta t \to 0} \frac{P[t_0 \leq T_n \leq t_0 + \Delta t]}{\Delta t}$$

$$= \lim_{\Delta t \to 0} \frac{1}{\Delta t} \int_{A(\Delta t)} \cdots \int \prod_i^n f(x_i) d\mathbf{x}.$$

Now

$$\sum_{i=1}^n \psi(x_i, u) = \sum_{i=1}^n \psi(x_i, t_0) + (t_0 - u) \sum_{i=1}^n D\psi(x_i, u(t_0))$$

where

$$t_0 \leq u(t_0) \leq u \leq t_0 + \Delta t.$$

Hence

$$f_n(t_0) = \lim_{\Delta t \to 0} \frac{c^{-n}(t_0)}{\Delta t} \int_{A(\Delta t)} \cdots \int \prod_1^n h_{t_0}(x_i) \exp\{\alpha(t_0)(t_0 - u) \sum_{i=1}^n D\psi(x_i, u(t_0))\} d\mathbf{x}$$

$$\leq c^{-n}(t_0) \lim_{\Delta t \to 0} \exp\{\alpha(t_0)(t_0 - u)k\} \lim_{\Delta t \to 0} \frac{1}{\Delta t} \int_{A(\Delta t)} \cdots \int \prod_1^n h_{t_0}(x_i) d\mathbf{x}$$

$$= c^{-n}(t_0) h_{t_0,n}(t_0).$$

Similarly $f_n(t_0) \geq c^{-n}(t_0) h_{t_0,n}(t_0)$ giving the required result. □

Using this result, the approximation to $f_n(t_0)$ can be obtained directly from an approximation to $h_{t_0,n}(t_0)$. It is important to note that for each point $t_0$, we use a different conjugate density. In the proof of the theorem, the property of $\alpha(t)$ specified in (4.3) has not been used, so that in fact the centering result (4.4) holds for arbitrary $\alpha$.

In order to approximate $h_{t_0,n}(t_0)$, it is necessary to express $T_n$ as an approximate mean and evaluate its cumulants. The development follows that of Field (1982) and Tingley (1987, p. 49-52). The first part is very similar to expansions used to demonstrate properties of maximum likelihood estimates. The result will be stated in terms of $\theta_0$, the true value of $\theta$ and then will be modified for the case of the conjugate density $h_{t_0}$ which is centered at $t_0$.

**Theorem 4.2** (Bhattacharya and Ghosh, 1978)

Assume that A4.2–A4.5 hold. Then there is a sequence of statistics $\{T_n\}$ such that for every compact $\mathbf{K}\subset\Theta$

$$\inf_{\theta_0\in K} P_{\theta_0}(|T_n-\theta_0| < d_0 n^{-1/2}(\log n)^{1/2}, T_n \text{ solves } (4.1))$$
$$= 1 - 0(1/\sqrt{n})$$

where $d_0$ may depend on $k$.

**Corollary**

With the above assumptions $|T_n-\theta_0|^m$ is $0_p(n^{-a})$ on every compact set $\mathbf{K}\subset\Theta$ provided that $m/2-a>0$.

The following construction enables us to write $T_n$ as an approximate mean. Consider the second order Taylor expansion of (4.1) about $\theta_0$:

$$\begin{aligned} 0 &= \frac{1}{n}\sum_{i=1}^{n}\psi(x_i,T_n) \\ &= \frac{1}{n}\sum_{i=1}^{n}[\psi(x_i,\theta_0)+D\psi(x_i,\theta_0)(T_n-\theta_0)+\frac{1}{2}D^2\psi(X_i,\theta_0)(T_n-\theta_0)^2] \\ &\quad + R_n(T_n) \end{aligned} \tag{4.5}$$

where $R_n(T_n)=0_p|T_n-\theta_0|^3=0_p(1/n)$.

Looking at the first three terms, let $Z=\psi(X,\theta_0)$, $Z_1=D\psi(X,\theta_0)$ and $Z_2= D^2\psi(X,\theta_0)$ and let $E_{\theta_0}Z_1=\mu_1$, $E_{\theta_0}Z_2=\mu_2$. Note that $E_{\theta_0}Z=0$.

Define

$$\mathbf{a}=(0,\mu_1,\mu_2)$$
$$\mathbf{Z}=(Z,Z_1,Z_2)$$

and let $q: R^4\to R$ be defined as

$$q(\mathbf{z},t)=z+(t-\theta_0)z_1+(t-\theta_0)^2z_2/2.$$

Note that since $q(\mathbf{a},\theta_0)=0$,we can apply the implicit function theorem to prove that there is a unique three times differentiable function $H: R^3\to R$ such that $H(\mathbf{a})=\theta_0$ and $q(\mathbf{z},H(\mathbf{z}))=0$ for $\mathbf{z}$ in a neighbourhood of $\mathbf{a}$. We now have, setting $\bar{\mathbf{z}}=(\bar{z},\bar{z}_1,\bar{z}_2)$:

**Lemma**

$$H(\bar{\mathbf{z}})-T_n=0_p(1/n).$$

Proof: cf. expression 2.39 and 2.40 of Bhattacharya and Ghosh (1978).

The next step is to expand $H$ in a Taylor series expansion about $\mathbf{a}$. The result is the following expression

$$H(\bar{\mathbf{z}})-\theta_0=-\bar{z}/A(\theta_0)+A(\theta_0)\mu_2\bar{z}^2/2+\bar{z}(\bar{z}_1-\mu_1)/A^2(\theta_0)+0_p(1/n).$$

Putting this equation together with the lemma, if A4.1–A4.5 hold, then for every $\theta_0$ in a compact subset of $\Theta$

$$T_n - \theta_0 = \bar{Z}/A(\theta_0) + A(\theta_0)\mu_2\bar{Z}^2/2 + \bar{Z}(\bar{Z} - \mu_1)/A^2(\theta_0) + 0_p(1/n). \tag{4.6}$$

In expression (4.6), all expectations are with respect to $\theta_0$ and are computed with the original density. If we now replace $\theta_0$ by $t_0$ and compute all expectations with respect to the conjugate density $h_{t_0}$, we have that $E_{t_0}Z = E_{t_0}\psi(X, t_0) = 0$ by the choice $\alpha$ in (4.3). Hence

$$T_n - t_0 = -\bar{Z}/A(t_0) + A(t_0)\mu_2\bar{Z}^2/2 + \bar{Z}(\bar{Z}_1 - \mu_1)/A^2(t_0) + 0_p(1/n). \tag{4.7}$$

In order to apply the Edgeworth expansion, we need to evaluate the cumulants of the left hand side of (4.7). For this purpose the results of James and Mayne (1962) are appropriate. Denote the cumulant of order $r$ of $n^{1/2}(T_n - t_0)$ by $\lambda^r$ and of $\bar{Z}$ and $\bar{Z}_1$ by $K^r$ and $K_1^r$. Since we are working with means it follows that $K^r$ and $K_1^r$ are of order $n^{-r+1}$. The cumulants of $n^{1/2}(T_n - t_0)$ can be expressed in terms of $K$ as follows: (cf. James and Mayne, 1962, p. 51).

$$\begin{aligned}
\lambda^1 &= n^{1/2}\{A(t_0)\mu_2 K^2/2 + K_1^1/A^2(t_0)\} + 0(n^{-3/2}) \\
&= d_1/n^{1/2} + 0(n^{-3/2}) \qquad (say) \\
\lambda^2 &= \sigma^2(t_0)/A^2(t_0) + 0(n^{-1}) \qquad \text{where} \qquad \sigma^2(t_0) = E_{t_0}[\psi^2(X, t_0)] \\
\lambda^3 &= d_3/n^{1/2} + 0(n^{-3/2}).
\end{aligned}$$

All higher order cumulants are of order $0(n^{-1})$ or higher .

The characteristic function of $n^{1/2}(T_n - t_0)$ can be written as

$$\varphi(u) = \exp\left\{\sum_{r=0}^{\infty} \lambda^r (iu)^r/r!\right\}$$

Using the above results we have

$$\varphi(u) = \exp\left\{d_1(iu)/n^{1/2} - \sigma^2(t_0)u^2/A^2(t_0) + d_3(iu)^3/3!n^{1/2} + 0(1/n)\right\}.$$

We use the result that

$$\int \exp\left\{-u^2\sigma^2(t_0)/A^2(t_0)\right\}\pi_r(iu)e^{-iux}du = \pi_r(-D)\phi_{t_0}(x)$$

where $\pi_r$ is a polynomial of order $r$, $D$ represents differentiation and $\phi_{t_0}$ is the normal density with mean 0 and variance $\sigma^2(t_0)/A^2(t_0)$. From this it follows that the density of $n^{1/2}(T_n - t_0)$ under $h_{t_0}$ at $x$ is given by

$$h(x) = \phi_{t_0}(x)\left[1 + dD(\phi_{t_0}(x))/n^{1/2} + d_3 D^{(3)}(\phi_{t_0}(x))/n^{1/2} + 0(1/n)\right].$$

For $x = 0$, the terms of order $n^{-1/2}$ drop out giving the density of $T_n$ at $t_0$ under $h_{t_0}$ as

$$n^{1/2}h(0) = n^{1/2}\phi_{t_0}(0)[1 + 0(1/n)].$$

Using the results of Theorem 4.1, it follows that

$$f_n(t_0) = (n/2\pi)^{1/2} c^{-n}(t_0)[A(t_0)/\sigma(t_0) + 0(1/n)].$$

To summarize:

**Theorem 4.3**

If $T_n$ represents the solution of $\sum_{i=1}^{n} \psi(x_i, t) = 0$ and A4.1–A4.5 hold, then an asymptotic expansion for the density of $T_n$ is

$$f_n(t_0) = (n/2\pi)^{1/2} c^{-n}(t_0) A(t_0)/\sigma(t_0)[1 + 0(1/n)] \tag{4.8}$$

where $\alpha(t_0)$ is the solution of $\int \psi(x, t_0) \exp\{\alpha\psi(x, t_0)\} f(x)dx = 0$, $c^{-1}(t_0) = \int \exp\{\alpha\psi(x, t_0)\} f(x)dx$, $\sigma^2(t_0) = E_{t_0}\psi^2(x, t_0)$, $A(t_0) = E_{t_0}[D\psi(x, t_0)]$ where $E_{t_0}$ is expectation with respect to the conjugate density

$$h_{t_0}(x) = c(t_0) \exp\{\alpha(t_0)\psi(x, t_0)\} f(x).$$

From a practical point of view, the error bound on the density in the theorem above is not of direct use. What is usually needed is an error on the integral of the density over a region of interest. For example in testing, we need to compute tail areas for the calculation of P-values. In order to begin, a slightly stronger version of theorem 4.3 is needed. Following the argument on p. 677 of Field (1982) it can be seen that the error bound in (4.8) is uniform for all $t$ in a compact set. We now use an argument very similar to that in Durbin (1980a, 310–316).

From the results on the cumulants of $n^{1/2}(T_n - \theta_0)$ we have that the fourth cumulant of $T_n - \theta_0$ is $0(n^{-3})$ and the variance is $0(n^{-1})$. This implies that the fourth moment is $0(n^{-2})$. We can now find a constant $C_1$, so that for $n$ sufficiently large, $E(|T_n - \theta_0|^4) < C_1 n^{-2}$. Letting $A$ be the region $|t - \theta_0| \leq d$, then

$$\begin{aligned} C_1 n^{-2} &> \int_A (t - \theta_0)^4 f_n(t)dt + \int_{A^c} (t - \theta_0)^4 f_n(t)dt \\ &> \delta^4 \int_A f_n(t)dt = \delta^4 P(|T_n - \theta_0| > \delta_2). \end{aligned}$$

This implies $P(|T_n - \theta_0| > \delta_2) = 0(n^{-2})$. Theorem 4.3 implies that

$$1 - C_2/n \leq f_n(t)/g_n(t) \leq 1 + C_2/n \quad \text{for} \quad |t - \theta_0| \leq \delta_2$$

where $g_n(t) = (n/2\pi)^{1/2} c^{-n}(t) A(t)/\sigma(t)$ and $C_2$ does not depend, on $n$ or $t$.

Let $\{B_n\}$ be a sequence of Borel sets such that $P(T_n \in B_n)$ converges to a positive limit.

Then

$$\begin{aligned} \left| \int_{B_n} (f_n(t) - g_n(t))dt \right| &\leq \left| \int_{A \cap B_n} f_n(t)dt - \int_{A \cap B_n} g_n(t)dt \right| + \left| \int_{A^c \cap B_n} f_n(t)\left(1 - \frac{g_n(t)}{f_n(t)}\right)dt \right| \\ &\leq 0(1/n^2) + \left| \int_A g_n(t)dt \right| + C_2/n. \end{aligned}$$

We will verify in section 6.2,

$$\int_{\delta_2+\theta_0}^{\infty} g_n(t)dt = 0(1/n).$$

Conditional on this last result, we can then conclude

$$\int_{\mathbf{B}_n} f_n(t)dt = \int_{\mathbf{B}_n} g_n(t)dt + 0(n^{-1}) \tag{4.9}$$

This result shows that $g_n(t)$ can be integrated with an error which is at most $0(n^{-1})$. In many inference problems, it is an approximation to the tail area which is of interest. From this point of view (4.9) is a much more useful result than (4.8).

In actually using (4.8) for computational purposes, it is recommended that the constant of integration be determined numerically. Some further discussion of this point is given in section 5.2. This technique gives an important increase in the accuracy of the approximation. In fact, Durbin (1980a) p. 317 gives a heuristic argument that the renormalization reduces the magnitude of the error from $n^{-1}$ to $n^{-3/2}$. Since we are using these approximations for very small $n$, often $n \leq 10$, any order terms on the error have to be keep in context. The usefulness of (4.8) and (4.9) is really determined by how well, they perform for small to moderate values of $n$. By the time, we have $n$ of 20 in many situations, the normal approximation may suffice.

We turn now to questions of computation of $g_n(t)$. To evaluate $g_n(t)$ at a specific point, the main computational effort is in computing $\alpha(t)$ from (4.3). This is a non-linear equation which involves a numerical integration for each function evaluation. For any cases where we have done computations, secant methods have proven very satisfactory. As an example, we computed $\alpha(t)$, $c(t)$, $\sigma^2(t)$ over a grid of 90 points for the mean with an underlying uniform distribution. The computations took 5.6 seconds of CPU time on a VAX 785 running 4.3 BSC Unix. In order to compute probabilities a simple way is to evaluate $g_n$ over a grid of points and then do numerical integration. If we are specifically interested in evaluating tail areas, very accurate integral approximations (uniform asymptotic expansions) are available which reduce the computational effort substantially. The basic reference is Lugannani and Rice (1980). In chapter 6, these ideas are developed more fully and the approximations to the tail area provided.

A useful feature of the computation is that once $\alpha(t)$, $c(t)$ and $\sigma(t)$ have been computed, we can evaluate $g_n(t)$ for any $n$. To give an indication of the accuracy of the approximation we consider the following situation.

The exact density for the Huber estimate with $\psi(x) = x$ if $|x| < k$, $= k$ sgn $x$ if $|x| \geq k$ has been calculated by P. Huber for two contaminated normal distributions and by A. Marazzi for the Cauchy distribution (unpublished). For the contaminated normal, the results are obtained by direct convolution and have been checked in double precision by D. Zwiers. The exact results for the Cauchy were calculated with fast Fourier transforms via characteristic functions. The effort in obtaining the exact results is substantially more than that required for the approximation and must be recomputed for each new value of $n$.

To measure, the accuracy of our approximation, the relative percent errors of tail areas are computed for selected values of $t$ for both the the contaminated normal and Cauchy. Note that the relative percent error for upper tails is computed as:

100 (approximate - exact cumulative)/(1− exact cumulative). The values are as given in Exhibit 4.1 and 4.2.

| t | n 1 | 2 | 3 | 4 | 5 | 6 | 7 | 8 | 9 |
|---|---|---|---|---|---|---|---|---|---|
| 0.1 | -0.05 | 0.12 | 0.05 | 0.04 | 0.04 | 0.03 | 0.03 | 0.03 | 0.03 |
| 0.5 | -0.70 | 0.90 | 0.40 | 0.40 | 0.30 | 0.30 | 0.30 | 0.20 | 0.20 |
| 1.0 | -5.40 | 3.70 | 0.40 | 1.00 | 0.70 | 0.70 | 0.70 | 0.60 | 0.70 |
| 1.5 | -23.20 | 13.50 | -3.60 | -0.05 | 1.50 | 1.50 | 1.10 | 1.10 | 0.90 |
| 2.0 | -60.80 | 30.50 | -24.00 | 19.80 | -10.20 | 12.50 | -4.00 | | |
| 2.5 | -99.60 | 41.60 | -64.60 | 37.80 | -46.70 | | | | |
| 3.0 | -116.70 | 45.40 | -90.20 | | | | | | |

**Exhibit 4.1a**

Percent relative errors of the Huber estimate ($k = 1.4$) for contaminated normal ($\epsilon = .05$) versus $t$.

| $100 \cdot (1 - G_n)$ | n 1 | 2 | 3 | 4 | 5 | 6 | 7 | 8 | 9 |
|---|---|---|---|---|---|---|---|---|---|
| 5 | -60.8 | 13.5 | 0.2 | 1.0 | 0.6 | 0.5 | 0.4 | 0.4 | 0.3 |
| 1 | | | -12.8 | 4.0 | 0.7 | 0.8 | 0.7 | 0.6 | 0.7 |
| 0.1 | | | | | -10.2 | 3.2 | 1.1 | 1.0 | 0.9 |
| 0.01 | | | | | | | -6.5 | 2.8 | 1.4 |

**Exhibit 4.1b**

Percent relative errors of the Huber estimate ($k = 1.4$) for contaminated normal ($\epsilon = .05$) versus approximate percentage points determined by $G_n$, the cumulative of $g_n$.

A glance at the relative errors for the contaminated normal shows that for $t$ values of 1 or less, the relative errors are all 1% or less even down to $n = 3$ and all remain under 10% (most under 3%) for $t = 1.5$. In terms of percentage points, the estimate is very accurate at the 1% level down to $n = 4$ with $\epsilon = 0.5$ and at the .1% level down to $n = 6$ with $\epsilon = .05$. It is only with small $n$ and large $t$ that the relative errors become larger and even here the estimate is fairly good. For instance with $n = 3$ and $t = 3.0$ (a relative error of 90%), the actual difference is .002(.99795 − .99610). The results when $\epsilon = .10$ are similar to those above.

| t | n 1 | 2 | 3 | 4 | 5 | 6 | 7 | 8 | 9 |
|---|---|---|---|---|---|---|---|---|---|
| 1 | -12.3 | 8.0 | -4.4 | 0.8 | -1.5 | 0.6 | -0.7 | -.03 | -0.5 |
| 3 | -21.0 | 23.3 | -12.6 | 14.1 | -7.0 | 8.5 | -4.0 | 4.7 | -2.6 |
| 5 | -33.6 | 33.6 | -24.9 | 24.9 | -16.2 | 18.6 | -12.2 | 13.0 | -7.3 |
| 7 | -43.5 | 40.3 | -37.2 | 33.1 | -28.0 | 27.8 | -16.7 | 22.5 | -16.7 |
| 9 | -51.2 | 44.8 | -47.8 | 38.6 | -37.5 | 35.7 | -29.8 | 31.0 | -16.7 |

**Exhibit 4.2**

Percent relative errors of the Huber estimate ($k = 1.4$) for the Cauchy.

For the Cauchy, the relative errors remain well under control even well out into the extreme tails. For instance with $n = 7$ and $t = 9$ (relative error of 30%), the actual difference is .00001(.99995 − .99994) so that the estimate would be very usable even at the .005% level. A graphical display of the results is given in Exhibit 4.3 and 4.4. The value of $\log(F/1-F)$ is plotted against $t$ for the 5% contaminated normal and the Cauchy. From the graphs it can be seen that in terms of critical values the results for the contaiminated normal with $\epsilon = .05$ are very accurate to $n = 3$ for the 5% level, to $n = 5$ at the 1% level, to $n = 7$ at the .1% level and to $n = 9$ (or even $n = 7$) at the .01% level. Similar results hold for the Cauchy. These results imply that the approximations accurately reflect the distributional behaviour of the estimates and provide a very useful tool for determining small sample properties of interest.

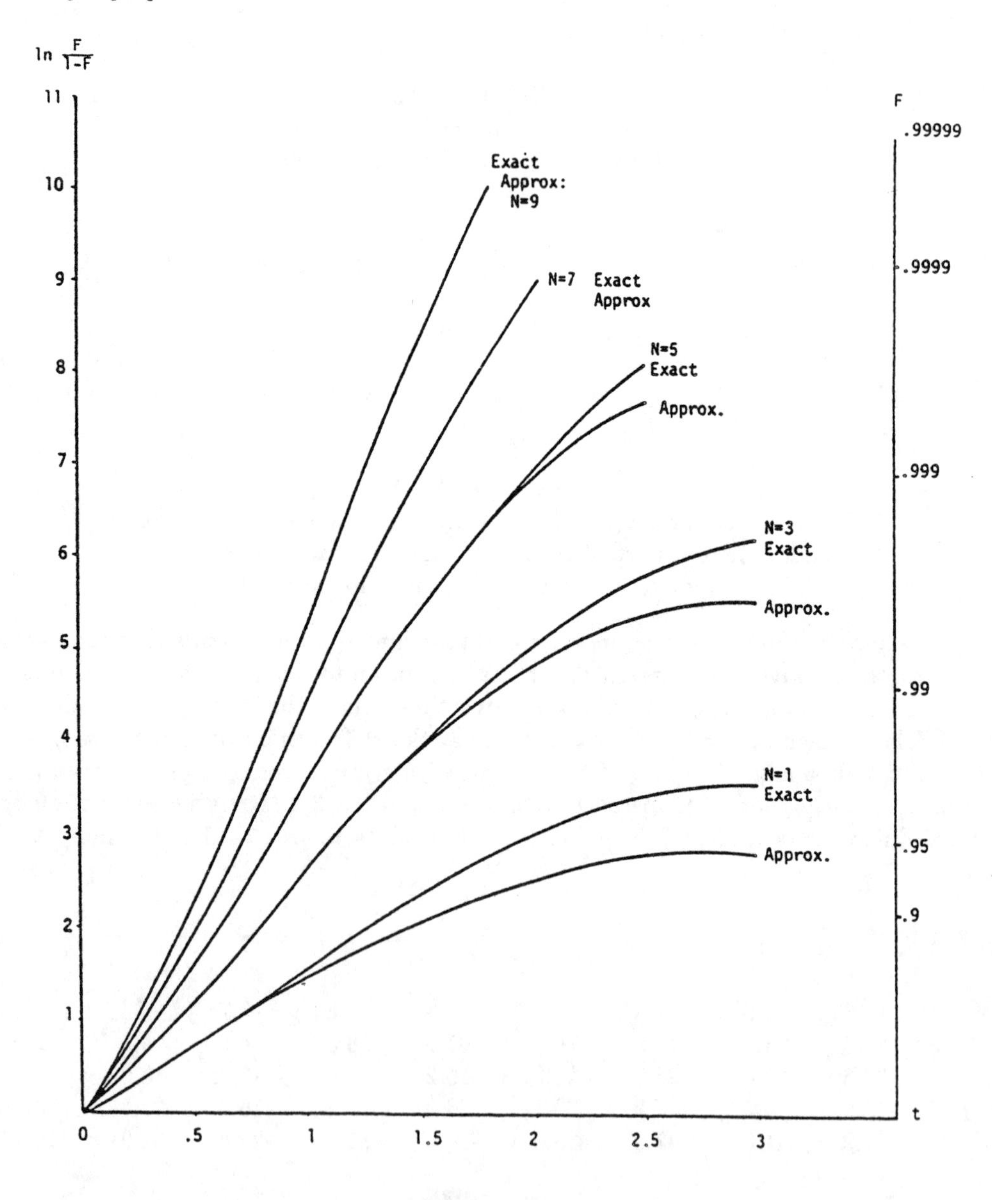


**Exhibit 4.3**
$\log(F_n/(1-F_n))$ versus $t$ of Huber estimate ($k = 1.5$) for contaminated normal with ($\epsilon = .05$)

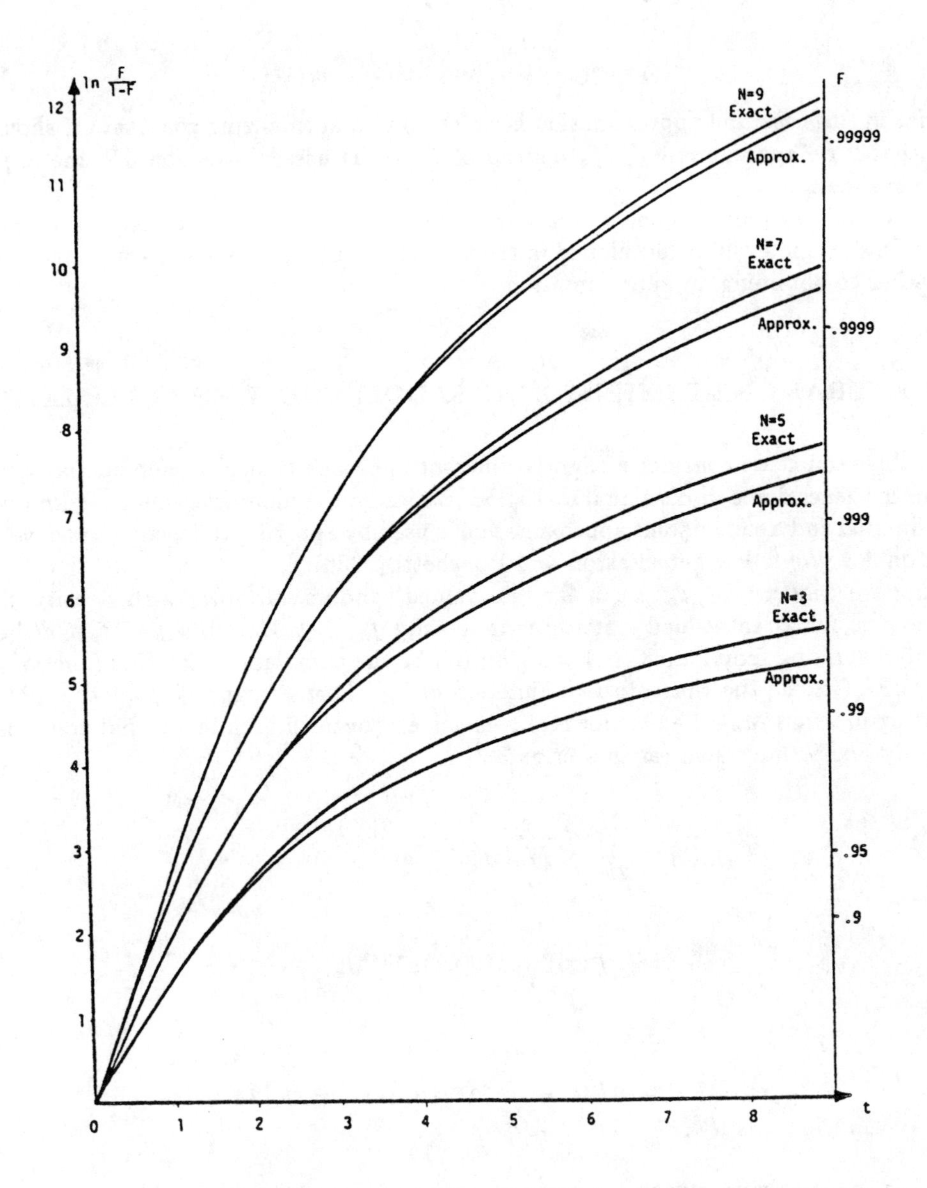


**Exhibit 4.4**

$\log(F_n/(1/F_n))$ versus $t$ of Huber estimate ($k = 1.5$) for Cauchy

Some calculations with the mean suggest that any accuracy obtained by including the first neglected term is of the order of round off errors and so the first term in the expansion is all that is needed for good accuracy.

As another example, consider the M-estimate version of $\beta$-quantiles with $\psi(x) = \beta - 1$ for $x < 0$, $\psi(x) = 0$ for $x = 0$ and $\psi(x) = \beta$ for $x > 0$. For those $n$ where the defining equation $\sum \psi(x_i - T_n) = 0$ has a unique solution, the exact density of the M-quantile is the density of the appropriate order statistic and is proportional to $(F(t))^{(n-1)\beta}(1 - F(t))^{(n-1)(\beta-1)} f(t)$. If we now compute the approximation given in (4.8), we find that

$$\alpha(t) = \log((1-\beta)F(t)/(\beta(1-F(t))),$$

$$c(t) = (1-\beta)\exp(-\beta\alpha(t))/(1-F(t)),$$

$$\sigma^2(t) = \beta(1-\beta) \quad \text{and} \quad A(t) = c(t)f(t)$$

and that in this case, the approximation is exact up to a normalizing constant. It should be noted that if we approximate $f_n'/f_n$ instead of $f_n$ as discussed in section 5.2, the approximation is exact.

As a final example, we could consider the example of logistic regression through the origin. This example will be developed in the next chapter where we compare several related approaches to obtaining an approximation.

## 4.3. GENERAL ONE-DIMENSIONAL SADDLEPOINT APPROXIMATIONS

In this section we consider a slightly different approach to derive saddlepoint approximations for general one-dimensional statistics. Basically, the moment generating function is approximated and a saddlepoint approximation is used by applying the techniques developed in section 3.3. We follow here Easton and Ronchetti (1986).

Suppose that $x_1, \cdots, x_n$ are n iid real valued random variables with density $f$ and $T_n(x_1, \cdots, x_n)$ is a real valued statistic with density $f_n$. Let $M_n(\alpha) = \int e^{\alpha t} f_n(t)dt$ be the moment-generating function, $K_n(\alpha) = \log M_n(\alpha)$ be the cumulant-generating function, and $\rho_n(\alpha) = M_n(i\alpha)$ be the characteristic function of $T_n$. Further suppose that the moment-generating function $M_n(\alpha)$ exists for real $\alpha$ in some nonvanishing interval that contains the origin. By Fourier inversion (as in section 3.3),

$$\begin{aligned} f_n(t) &= \frac{1}{2\pi} \int_{-\infty}^{+\infty} M_n(ir)e^{-irt}dr \\ &= (n/2\pi i) \int_{\mathcal{I}} M_n(nz)e^{-nzt}dz \\ &= (n/2\pi i) \int_{\tau-i\infty}^{\tau+i\infty} \exp\{n[R_n(z) - zt]\}dz, \end{aligned} \tag{4.10}$$

where $\mathcal{I}$ is the imaginary axis in the complex plane and $\tau$ is any real number in the interval where the moment generating function exists, and

$$R_n(z) = K_n(nz)/n \ . \tag{4.11}$$

Note that if $T_n$ is the arithmetic mean, then $R_n(z) = K(z)$, the cumulant generating function of the underlying density $f$, and in this case (4.10) equals formula (3.6). The problem in (4.10) is that in general $R_n$ is *unknown*. The basic idea of the general saddlepoint approximation is to approximate $R_n(z)$ and then apply the saddlepoint technique to the integral in (4.10).

For a given problem, one can find sometimes special approximations to the cumulant generating function and to $R_n$. In very special situations these functions are even known exactly. However, since the goal of this approach is to derive a saddlepoint approximation for a *general* model, we do not want to assume any special structure in the problem. In this framework, classical Edgeworth expansions provide a very important source of general approximations to $\log \rho_n(\alpha)$ and therefore to $R_n(z)$. Moreover, they have the advantage of being good local approximations around the center of the distribution. This property is exploited by the saddlepoint approximation which can be viewed as a low order Edgeworth approximation locally at each point (see section 3.4 and Remark 4.4 below). For these reasons we will use the Edgeworth approximation as a basic construction tool and will show that whenever an Edgeworth expansion for the density $f_n$ of $T_n$ is available, a saddlepoint approximation can be carrried out that will generally improve it.

Let us now work out the Edgeworth approximation for $R_n(z)$. Denote by $\tilde{f}_n$ the Edgeworth expansion for $f_n$ up to and including the term of order $n^{-1}$. Let $\tilde{M}_n$ and $\tilde{K}_n$ be the moment generating function and the cumulant generating function of $\tilde{f}_n$, respectively. (Technically, $\tilde{M}_n$ and $\tilde{K}_n$ may not be moment generating and cumulant generating functions, since $\tilde{f}_n$ may not be a density. We will continue to use this terminology, however.) Let $\tilde{R}_n(z) = \tilde{K}_n(nz)/n$. Then, from the Edgeworth approximation (up to the term of order $n^{-1}$) one can obtain an approximation for $\log \rho_n(\alpha)$ and, therefore, $R_n(z)$ in terms of the first four cumulants. That is, $R_n(z)$ can be approximated by

$$\tilde{R}_n(z) = \mu_n z + \frac{n\sigma_n^2 z^2}{2} + \frac{\kappa_{3n}\sigma_n^3 n^2 z^3}{6} + \frac{\kappa_{4n}\sigma_n^4 n^3 z^4}{24}, \tag{4.12}$$

where $\mu_n$ is the mean, $\sigma_n^2$ the variance of $T_n$ and $\kappa_{jn}$ are the cumulants of $T_n/\sigma_n$. Note that $\mu_n = 0(1)$, $\sigma_n = 0(n^{-1/2})$, and $\kappa_{jn} = 0(n^{-j/2+1})$ for $j = 3, 4$, since we have assumed that the Edgeworth expansion up to and including the term of order $n^{-1}$ for $f_n$ exists. In general, $\mu_n$, $\sigma_n$ and $\kappa_{jn}$ are not known exactly, but expansions up to the appropriate order will suffice to keep the same order in the approximation. Applying the saddlepoint technique to the integral in (4.21) gives the saddlepoint approximation of $f_n$ with uniform error of order $n^{-1}$:

$$g_n(t) = \left[\frac{n}{2\pi\tilde{R}_n''(\alpha_0)}\right]^{1/2} \exp\{n[\tilde{R}_n(\alpha_0) - \alpha_0 t]\}, \tag{4.13}$$

where $\alpha_0$ is the saddlepoint determined as a root of the equation

$$\tilde{R}_n'(\alpha_0) = t, \tag{4.14}$$

$\tilde{R}_n(\alpha) = \tilde{K}_n(n\alpha)/n$ is given by (4.12) and $\tilde{R}_n'$ and $\tilde{R}_n''$ denote the first two derivatives of $\tilde{R}_n$.

As usual the approximation $g_n(t)$ given by (4.13) can be improved by renormalization, that is by computing numerically $C_n = \int_{-\infty}^{+\infty} g_n(t)dt$ to obtain $g_n(t)/C_n$ which integrates to 1. Tail areas can be obtained from (4.13) by numerical integration. However, sometimes it is convenient to have a direct saddlepoint approximation for the tail area $P[T_n > a]$. By the same argument as in Daniels (1983) (see section 6.2), the saddlepoint equation (4.14) can be used as a change of variables to obtain the approximation

$$P(T_n > a] \cong \int_{\alpha_\ell}^{\infty} \left[\frac{n\tilde{R}_n''(\alpha_0)}{2\pi}\right]^{1/2} \exp\{n[\tilde{R}_n(\alpha_0) - \alpha_0\tilde{R}_n'(\alpha_0)]\}d\alpha_0, \tag{4.15}$$

where $\tilde{R}_n'(\alpha_\ell) = a$.

*Remark 4.1*
In the case where the cumulant generating function $K_n$ is known exactly, one can apply directly the saddlepoint method to (4.10). In this case, the saddlepoint approximation is given by (4.13) and (4.14), with $\tilde{R}_n$ replaced by the exact $R_n(\alpha) = K_n(n\alpha)/n$.

In order to show a connection with a result by Chaganty and Sethuraman (1985), we now compute the approximation (4.13) for the statistic $T_n/n$ and *in the case where $K_n$ is known exactly*. Let us denote by $f_n$, $g_n$, $K_n$, $R_n$ the functions for $T_n$ and by $f_n^*$, $g_n^*$, $K_n^*$, $R_n^*$ the corresponding functions for $T_n/n$. Then we have

$$R_n^*(\alpha) = K_n(\alpha)/n,$$
$$R_n^{*\prime}(\alpha) = K_n'(\alpha)/n,$$
$$R_n^{*\prime\prime}(\alpha) = K_n''(\alpha)/n,$$

and the saddlepoint approximation for $T_n/n$ evaluated at a point $t_n$

$$g_n^*(t_n) = \left[\frac{n}{2\pi R_n^{*\prime\prime}(\alpha_n)}\right]^{1/2} \exp\{n[R_n^*(\alpha_n) - \alpha_n t_n]\}, \tag{4.16}$$

where $\alpha_n$ is determined by the equation

$$R_n^{*\prime}(\alpha_n) = t_n.$$

Equation (4.16) is exactly equation (2.1) of Theorem 2.1 in Chaganty and Sethuraman (1985). The same authors generalize the approximation (4.16) to an arbitrary multidimensional statistics under the condition that the cumulant generating function is known; see Chaganty and Sethuraman (1986).

*Remark 4.2*
Since $\tilde{R}_n'(\alpha)$ is a third degree polynomial, the solution of equation (4.14) does not pose any computational problems. However, this equation can have multiple real solutions and only the solution $\alpha_0$ with $\tilde{R}_n''(\alpha_0) > 0$ can be used for the saddlepoint approximation. A simple example in which this happens is the approximation to the density of $s^2 = (n-1)^{-1}\sum_{i=1}^n (x_i - \bar{x})^2$ where $x_1, \cdots, x_n$ are $n$ iid observations from a $N(0,1)$.

*Remark 4.3*
Another possible approximation for $R_n(z)$ can be obtained using $\tilde{p}_n(\alpha)$ as given by the Edgeworth approximation instead of the expansion of $\log \tilde{p}_n(\alpha)$. This amounts to approximating $R_n(z)$ by

$$\tilde{R}_n(z) = \mu_n z + n\sigma_n^2 z^2/2 + \frac{1}{2}\log\left(1 + \frac{n^3\sigma_n^3 z^3 \kappa_{3n}}{6} + \frac{3\kappa_{4n} n^4 \sigma_n^4 z^4 + \kappa_3^2 n^6 \sigma_n^6 z^6}{72}\right).$$

We have no numerical experience with this approach.

*Remark 4.4*
One can use the same kind of computations as in section 4.2 to express $f_n$ by means of its conjugate density, namely,

$$f_n(t) = e^{n(R_n(\tau) - \tau t)} h_{\tau,n(t)}$$

where $h_{\tau,n}(t)$ is the density of $T_n$ with the underlying conjugate density. The choice $\tau = \alpha_0$ and an Edgeworth expansion of $h_{\tau,n}(t)$ leads to the saddlepoint approximation of $f_n$. This approach also requires knowledge of the exact cumulant-generating function. The general saddlepoint approximation described earlier corresponds to using the Edgeworth expansion to obtain an approximation to the conjugate density. Thus,

$$f_n(t) = e^{n(\tilde{R}_n(\tau)-\tau t)}\tilde{h}_{\tau,n}(t) + D_n(t)$$

where $\tilde{R}_n(z)$ is given by (4.12), $\tilde{h}_{\tau,n}(u)$ is the conjugate density of $\tilde{f}_n$, and $D_n(t) = f_n(t) - \tilde{f}_n(t)$. Note that the term of order $n^{-1/2}$ disappears because $\tilde{f}_n$ is recentered at $t$ through $\tilde{h}_{\tau,n}$; that is,

$$\tilde{R}_n'(\tau) - t = 0$$

if $\tau = \alpha_0$, the saddlepoint.

From the conjugate density point of view, the development of this approach is similar in spirit to that of Durbin (1980a) and Barndorff-Nielsen (1983), (see section 5.3), but we do not restrict ourselves to sufficient statistics or to maximum likelihood estimators nor do we assume any underlying parametric model. It should be noted, however, that when a special structure exists, saddlepoint approximations that have been developed to exploit this structure should be used, as they will generally perform better than this general approach.

*Remark 4.5*
As was pointed out by J. W. Tukey and B. Efron, this technique could be applied iteratively as follows. Start with an approximation $\tilde{R}_n^{(1)}$ for $R_n$ (given for instance by an Edgeworth approximation) and apply the saddlepoint technique to (4.10) to get the saddlepoint approximation $g_n$ in (4.13). Now, by numerical integration, compute a new approximation for $R_n$,

$$\tilde{R}_n^{(2)}(z) = (1/n)\log \tilde{M}_n^{(2)}(nz),$$

where $\tilde{M}_n^{(2)}(\alpha) = \int e^{\alpha t} g_n(t)dt$, then compute a new saddlepoint approximation for the density. This can be repeated until convergence is reached. This iteration process may improve the original approximation but its performance is an open question.

*Remark 4.6*
An alternative way of approximating the density of a general statistic by means of saddlepoint techniques is the following (see Field 1982 and section 4.5). Suppose $T_n$ can be written as a functional $T$ of the empirical distribution function $F^{(n)}$; that is, $T_n = T(F^{(n)})$. First, linearize $T_n$ using the first term of a von Mises expansion (see von Mises 1947),

$$T_n \cong T(F) + L_n(T,F), \tag{4.17}$$

where $F$ is the underlying distribution of the observations,

$$L_n(T,F) = \frac{1}{n}\sum_{i=1}^{n} IF(x_i;T,F), \tag{4.18}$$

and $IF(x;T,F)$ is the influence function of $T$ at $F$ (cf. Hampel 1968, 1974 and section 2.5). Then apply the classical saddlepoint approximation to $L_n(T,F)$, which is just an average of iid random variables. For instance, Tingley and Field (1988) apply the Lugannani and Rice approximation for the tail areas to (4.18); see section 6.3.

## 4.4. L-ESTIMATORS

In this section we apply the general saddlepoint technique to derive approximations to the density of linear combinations of order statistics.

We consider statistics of the form

$$T_n = \frac{1}{n} \sum_{i=1}^{n} c_{in} x_{(i)}, \tag{4.19}$$

where $x_{(1)} \leq x_{(2)} \leq \cdots \leq x_{(n)}$ are the order statistics and $c_{1n}, \cdots, c_{nn}$ are weights generated by a function $J : (0,1) \to R$,

$$c_{in} = J[i/(n+1)], \qquad i = 1, \cdots, n.$$

Typically the conditions imposed on $J$ are those that guarantee the existence of an Edgeworth expansion.

The distribution properties of L-statistics have been investigated by many authors. Exact distributions under special underlying distributions can be found in Weisberg (1971), Cicchitelli (1976), and David (1981). Asymptotic normality of these statistics has been shown under different sets of conditions (e.g., see Chernoff, Gastwirth and Johns, 1967; Shorack 1969, 1972; Stigler, 1969, 1974; David, 1981). Parr and Schucany (1982) investigated the small sample behavior of L-estimators via jackknifing. Finally, Helmers (1979, 1980) and van Zwet (1979) derived Edgeworth expansions for L-statistics with remainder $0(n^{-3/2})$. These will be the basic elements of our approximation, which we use in conjunction with the saddlepoint technique. In this section we summarize the numerical results obtained via saddlepoint approximation by Easton and Ronchetti (1986) for two L-estimators. These results show the great accuracy of this approximation down to very small sample sizes. These techniques can be used to approximate the distribution of more general L-statistics. One possible application is to so-called broadened letter values or "bletter values" (means of blocks of order statistics) suggested by Tukey (1977) as an improvement of the usual letter values in exploratory data analysis.

### 4.4.a The Asymptotically Most Efficient L-Estimator Under The Logistic Distribution

In this example we consider the asymptotically first-order efficient L-estimator for the center $\theta$ of the logistic distribution

$$F(x - \theta) = 1/[1 + \exp(-(x - \theta))].$$

This L-estimator is of the form (4.19) with the weight function $J(s) = 6s(1-s)$ and weights $c_{in} = 6i(1 - i/(n+1))/(n+1)$. We apply the technique presented in 4.3 to compute an approximation to the distribution of the statistic $n^{1/2}(T_n - \mu)/\sigma$, where $\mu(= 0)$ and $\sigma$ are the asymptotic mean and variance of $T_n$ under the logistic $F$. The Edgeworth expansion required in our formula is taken from Helmers (1980). It should be noted that the third moment equals 0 because of symmetry, so the term of order $n^{-1/2}$ disappears in the Edgeworth expansion. Thus the latter is of order $n^{-1}$ and should be very competitive with the saddlepoint approximation.

Numerical results for the cumulative distribution for sample sizes $3, 4, 10$ are given in Exhibits 4.5, 4.6, 4.7 for the right half of the distribution, since the density is symmetric.

The exact values are taken from Helmers (1980) who computed them by numerical integration for sample sizes 3 and 4, and by Monte Carlo simulation for sample size 10. The saddlepoint approximation for the cumulative is obtained by numerical integration from the saddlepoint approximation for the density computed using (4.13) and (4.14) for about 500 $t$ values. Exhibit 4.8 shows the residuals from the exact density for the rescaled saddlepoint, Edgeworth, and normal approximations. This plot clearly indicates that the rescaled saddlepoint approximation overall improves the Edgeworth approximation.

| x | Exact | Rescaled Saddlepoint | Unscaled Saddlepoint | Edgeworth | Normal |
|---|---|---|---|---|---|
| 2 | .5640 | .5617 | .5735 | .5536 | .5793 |
| .4 | .6262 | .6217 | .6320 | .6069 | .6554 |
| .6 | .6850 | .6787 | .6874 | .6592 | .7257 |
| .8 | .7391 | .7314 | .7387 | .7099 | .7881 |
| 1.0 | .7875 | .7790 | .7850 | .7582 | .8413 |
| 1.2 | .8248 | .8210 | .8259 | .8032 | .8849 |
| 1.4 | .8658 | .8572 | .8610 | .8439 | .9192 |
| 1.6 | .8958 | .8877 | .8908 | .8796 | .9452 |
| 1.8 | .9202 | .9130 | .9154 | .9100 | .9641 |
| 2.0 | .9397 | .9335 | .9353 | .9348 | .9772 |
| 2.2 | .9550 | .9499 | .9513 | .9543 | .9861 |
| 2.4 | .9669 | .9628 | .9638 | .9691 | .9918 |
| 2.6 | .9758 | .9727 | .9734 | .9798 | .9953 |
| 2.8 | .9825 | .9802 | .9807 | .9873 | .9974 |
| 3.0 | .9875 | .9858 | .9862 | .9923 | .9987 |

**Exhibit 4.5**

Exact cumulative distribution and approximations for sample size 3 for the asymptotically best L-estimator under logistic distribution.

| x | Exact | Rescaled Saddlepoint | Unscaled Saddlepoint | Edgeworth | Normal |
|---|---|---|---|---|---|
| .2 | .5663 | .5650 | .5750 | .5601 | .5793 |
| .4 | .6307 | .6281 | .6366 | .6190 | .6554 |
| .6 | .6919 | .6877 | .6949 | .6758 | .7257 |
| .8 | .7469 | .7424 | .7484 | .7295 | .7881 |
| 1.0 | .7963 | .7914 | .7962 | .7790 | .8413 |
| 1.2 | .8391 | .8341 | .8379 | .8236 | .8849 |
| 1.4 | .8752 | .8703 | .8732 | .8627 | .9192 |
| 1.6 | .9049 | .9003 | .9026 | .8960 | .9452 |
| 1.8 | .9287 | .9247 | .9264 | .9235 | .9641 |
| 2.0 | .9474 | .9440 | .9453 | .9454 | .9772 |
| 2.2 | .9618 | .9591 | .9600 | .9622 | .9861 |
| 2.4 | .9726 | .9705 | .9712 | .9748 | .9918 |
| 2.6 | .9807 | .9791 | .9796 | .9837 | .9953 |
| 2.8 | .9865 | .9854 | .9857 | .9898 | .9974 |
| 3.0 | .9907 | .9899 | .9902 | .9939 | .9987 |

**Exhibit 4.6**

Exact cumulative distribution and approximations for sample size 4 for the asymptotically best L-estimator under the logistic distribution.

| x | Exact | Rescaled Saddlepoint | Unscaled Saddlepoint | Edgeworth | Normal |
|---|---|---|---|---|---|
| .2 | .5734 | .5725 | .5776 | .5716 | .5793 |
| .4 | .6445 | .6426 | .6468 | .6409 | .6554 |
| .6 | .7089 | .7080 | .7115 | .7058 | .7257 |
| .8 | .7680 | .7670 | .7698 | .7647 | .7881 |
| 1.0 | .8196 | .8186 | .8208 | .8164 | .8413 |
| 1.2 | .8629 | .8622 | .8638 | .8604 | .8849 |
| 1.4 | .8985 | .8978 | .8990 | .8966 | .9192 |
| 1.6 | .9275 | .9260 | .9269 | .9255 | .9452 |
| 1.8 | .9486 | .9477 | .9483 | .9478 | .9641 |
| 2.0 | .9646 | .9639 | .9644 | .9645 | .9772 |
| 2.2 | .9764 | .9757 | .9760 | .9766 | .9861 |
| 2.4 | .9845 | .9840 | .9842 | .9850 | .9918 |
| 2.6 | .9905 | .9897 | .9898 | .9907 | .9953 |
| 2.8 | .9937 | .9935 | .9936 | .9944 | .9974 |
| 3.0 | .9959 | .9960 | .9961 | .9968 | .9987 |

**Exhibit 4.7**
Exact cumulative distribution and approximations for sample size 10 for the asymptotically best L-estimator under logistic distribution.

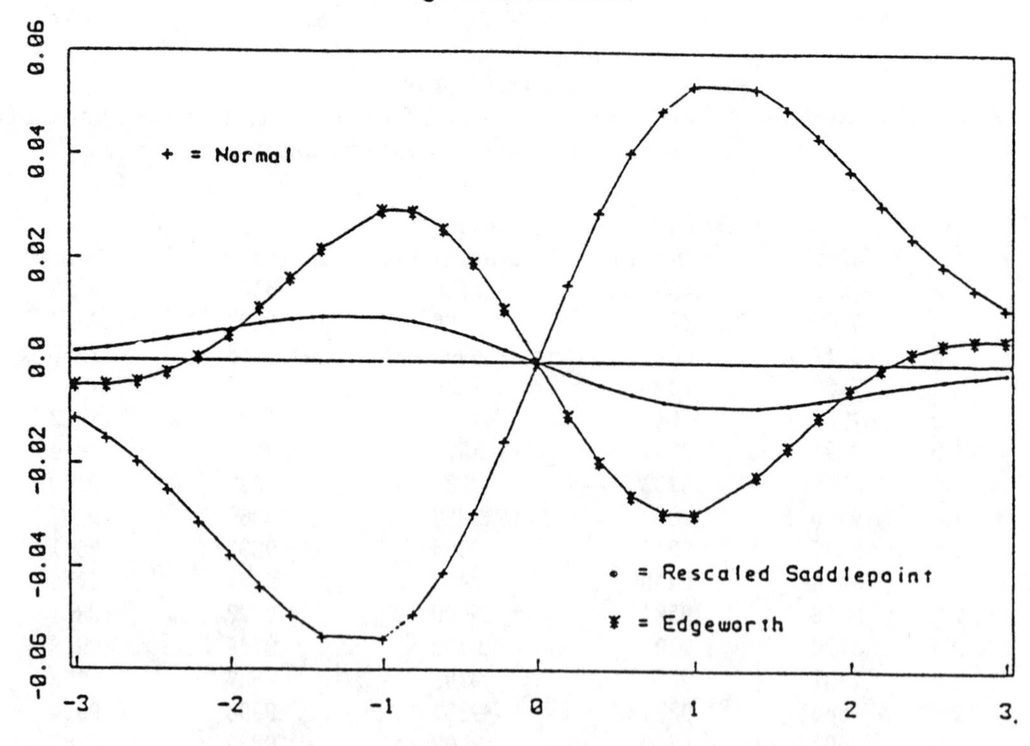


**Exhibit 4.8**
Error of the approximations of the density for sample size 3 for the asymptotically best L-estimator under logistic distribution.

In addition, unlike the Edgeworth and normal approximations, the Exhibits show that the rescaled saddlepoint approximation is wider tailed than the exact distribution, so its error is in the direction of giving conservative tests and confidence intervals. The same pattern can be seen for sample size 4 (not shown).

Although we are not approximating the distribution function directly, in practice these approximations may be used for calculating tail areas. Thus it is of interest to see how the saddlepoint approximation performs in the tails. Exhibit 4.9 shows the right-tail probability error for the right half of the distribution (for sample size 3) for the unscaled saddlepoint, rescaled saddlepoint, Edgeworth and normal approximation. The same pattern can be seen for sample size 10 (not shown).

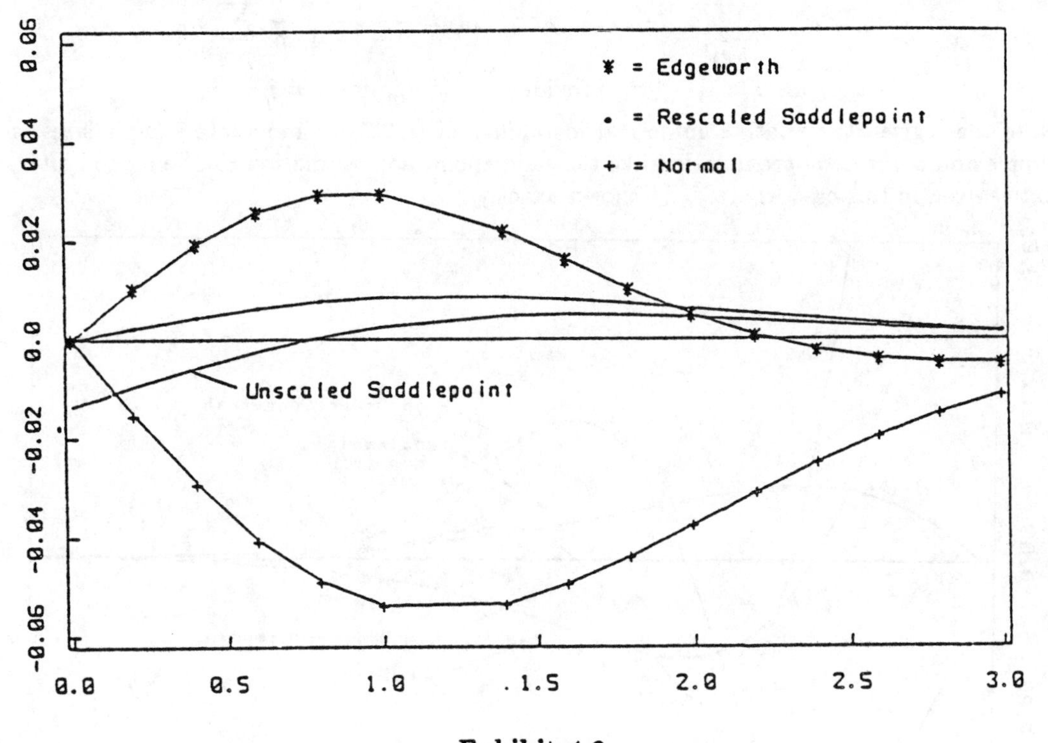


**Exhibit 4.9**

Error for the tail area for sample size 3 (cf. Exhibit 4.5).

Overall it appears that the rescaled saddlepoint technique generally improves on the Edgeworth approximation with respect to tail area and tends to err in the direction which produce conservative tests and confidence intervals.

### 4.4.b Trimmed Means of Exponential Observations

This example considers approximations to the distribution of trimmed means of exponential observations. Let $\alpha_\ell$ and $\alpha_u$ be the fraction of the observations trimmed from the upper and lower tails, respectively. We consider statistics of the form (4.19), where

$$\begin{aligned} c_{in} &= 0 \qquad \text{for } i \leq n\alpha_\ell \quad \text{or} \quad i \geq n(1-\alpha_u) \\ &= n/k \qquad \text{otherwise,} \end{aligned}$$

where $k$ is the number of nonzero weights. Note that $(1/n)\sum_{i=1}^{n} c_{in} = 1$.

Helmers (1979) derived the Edgeworth expansion for the distribution of $(T_n - \mu_n)/\sigma_n$, for trimmed linear combinations of order statistics with general weights on the observations between the $\alpha_\ell$ and $1-\alpha_u$ sample quantiles and with zero weights on the remaining observations. This expansion forms the basis for our general saddlepoint approximation.

The density of certain linear combinations of exponential order statistics can be written explicitly (see David, 1981). In our case the exact density of $T_n$ is

$$f_n(t) = \sum_{i=1}^{n} \frac{w_{in}}{a_{in}} \exp\left(-\frac{t}{a_{in}}\right), \tag{4.20}$$

where

$$w_{in} = a_{in}^{n-1} / \prod_{h \neq i} (a_{in} - a_{hn}) \quad \text{and} \quad a_{in} = \frac{1}{n-i+1} \frac{1}{n} \sum_{j=i}^{n} c_{jn}$$

$$\text{for } i = 1, \cdots, n, \text{ provided } a_{in} \neq a_{jn} \text{ for } i \neq j.$$

Note that, given the relative numerical instability of (4.20) for moderate sample sizes, a simple and accurate approximation like the saddlepoint approximation can be a good alternative even in this case where $f_n$ is known exactly .

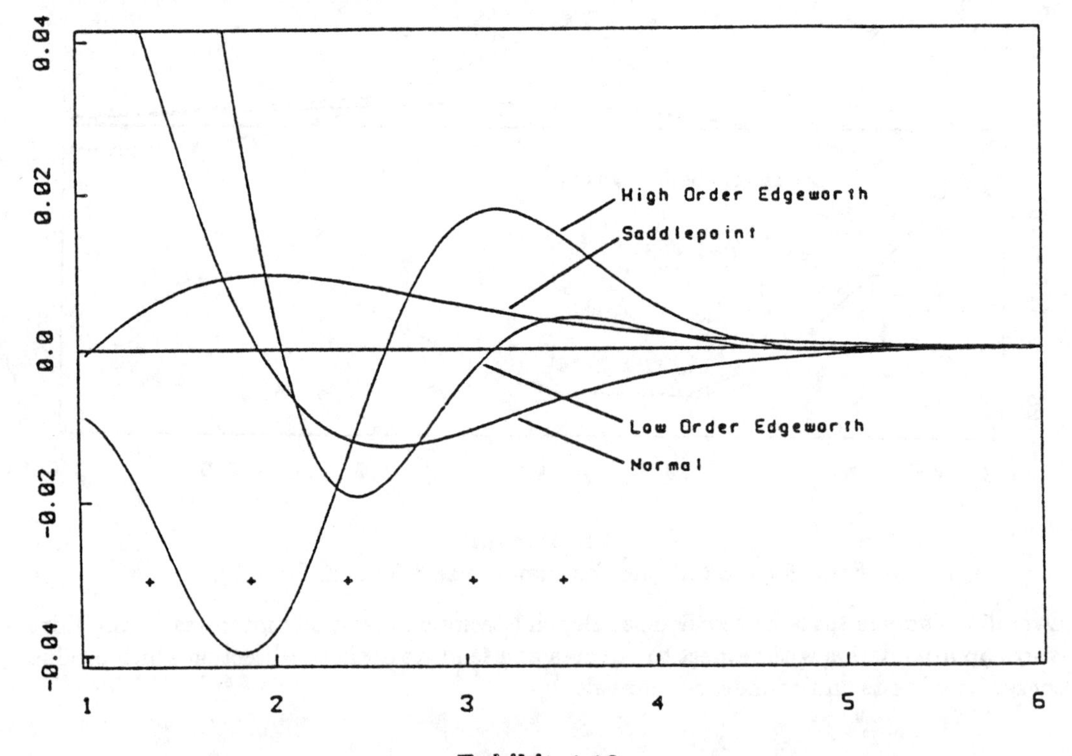


**Exhibit 4.10**

Error from the exact density in the right tail for 20% trimmed mean of 5 exponential observations. High (low) Edgeworth includes terms to order $n^{-1}(n^{-1/2})$. + marks denote .90, .95, .975, .99, .995 quantiles of exact distribution.

The figure shows that the saddlepoint approximation tends to be fairly stable and generally slightly wide throughout the tail. Both of the Edgeworth approximations show polynomial-like waves (and also become negative). The low order Edgeworth crosses the exact distribution a couple of times in the tail switching from being too wide to too narrow and back. The low-order Edgeworth approximation performs much better than the high-order Edgeworth approximation throughout this region and is competitive with the saddlepoint approximation in the 5% tail. It is sometimes too narrow, however. In the 5% tail, the error in the normal approximation is only slightly larger in absolute value than the error in the saddlepoint approximation, but the normal approximation is uniformly narrow.

Exhibit 4.11 plots the error in the approximate tail areas for the right 10% tail. As in the density case, the error in the Edgeworth approximations shows wavy behavior whereas the saddlepont approximation is uniformly wide.

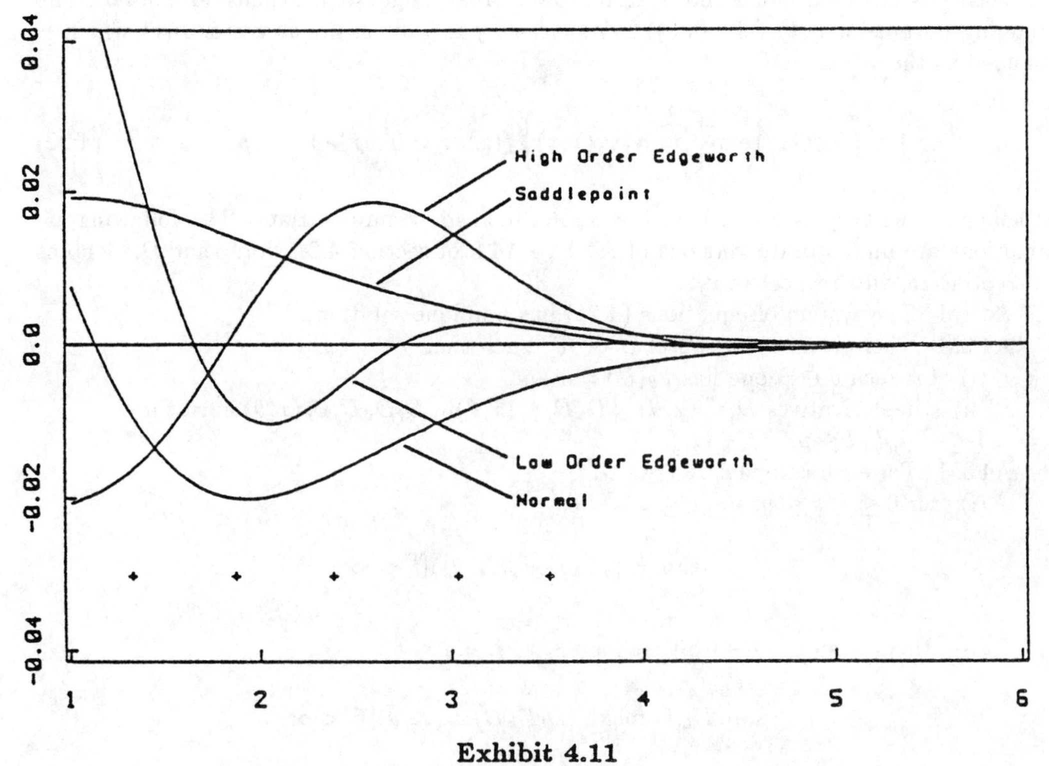


**Exhibit 4.11**
Error from the exact distribution in the right tail for the 20% trimmed mean of 5 exponential observations.

This example shows once more that the saddlepoint approximation exhibits some desirable properties that the Edgeworth approximations do not have. First, the saddlepoint approximation is unimodal and does not show the polynomial-like waves exhibited by the Edgeworth approximations. Thus the error in the saddlepoint approximation tends to be stable locally. Finally, the saddlepoint approximation tends to be wide in the tails so that error is in the direction of giving conservative tests and confidence intervals.

## 4.5. MULTIVARIATE M-ESTIMATORS

### 4.5.a Approximation

We return to the case of M-estimates as in section 4.2 but consider now the multiparameter problem. The M-estimate $\mathbf{T}_n$ for $\theta = (\theta_1, \theta_2, \cdots, \theta_p)$ is the solution of the system of equations:

$$\sum_{i=1}^{n} \psi_j(x_i, \mathbf{t}) = 0 \quad \text{for} \quad j = 1, \cdots, p. \tag{4.21}$$

The problem, as before, is to approximate the density $f_n(\mathbf{t})$ of $\mathbf{T}_n$. The development is similar to that presented in section 4.2 in that $\mathbf{T}_n$ is expanded as a multivariate mean, the density is centered and a multivariate lower order Edgeworth expansion is used. The centering parameter $\alpha(\mathbf{t})$ (cf. (4.3)) is replaced by a p-dimensional vector $\alpha(\mathbf{t})$ which is obtained as the solution of

$$\int \psi_j(x, \mathbf{t}) \exp\left\{\sum_{j=1}^{p} \alpha_j \psi_j(x, \mathbf{t})\right\} f(x) dx = 0 \quad j - 1, \cdots, p. \tag{4.22}$$

Although we write $x$ as univariate, the results hold for $\mathbf{x}$ multivariate. The following assumptions are multivariate versions of A4.1 to A4.5 of section 4.2. Note that $D_j$ denotes differentiation with respect to $\theta_j$.

A4.1M The system of equations (4.21) has a unique solution.

A4.2M There is an open subset $U$ of $R^m$ such that

(i) for each $\theta \in \Theta$ one has $F_\theta(U) = 1$ and

(ii) the derivatives $D_j\psi_r(x, \theta)$, $D_k D_j \psi_r(x, \theta)$, $D_l D_k D_j \psi_r(x, \theta)$ exist for $1 \leq r, j, k, l \leq p$.

A4.3M For each compact $K \subset \Theta$,

(i) for $0 \leq j, k \leq p$, $1 \leq r \leq p$,

$$\sup_{\theta_0 \in K} E_{\theta_0} |D_k D_j \psi_r(X, \theta_0)|^4 < \infty.$$

(ii) there is an $\epsilon > 0$ such that for $1 \leq r, j, k, l \leq p$,

$$\sup_{\theta_0 \in K} E_{\theta_0} \Big( \max_{|\theta - \theta_0| < \epsilon} |D_l D_k D_j \psi_r(X, \theta)| \Big)^3 < \infty.$$

A4.4M For each $\theta_0 \in \Theta$

$$E_{\theta_0} \psi_r(X, \theta_0) = 0$$

and the matrices

$$A(\theta_0) = E_{\theta_0} \left[ \frac{\partial \psi}{\partial \theta}(X, \theta_0) \right]$$

$$C(\theta_0) = E_{\theta_0} \left[ \psi(X, \theta_0) \psi^T(X, \theta_0) \right]$$

are non singular.

A4.5M The functions $A(\theta)$ and $E_\theta[(D_{k_1} D_{j_1} \psi_{r_1})(D_{k_2} D_{j_2} \psi_{r_2})]$, $0 \leq j_1, j_2, k_1, k_2 \leq p$, $k_1 + j_1 \geq 1$, $k_2 + j_2 \geq 1$, $1 \leq r_1, r_2 \leq p$, are continuous on $\Theta$.

At this point, a multivariate centering result is required. For a fixed $\mathbf{t}$, the conjugate density is

$$h_{\mathbf{t}}(x, \theta) = c(\mathbf{t}) \exp\left\{\sum_{j=1}^{p} \alpha_j \psi_j(x, \mathbf{t})\right\} f(x, \theta).$$

**Theorem 4.4**

Assume that the joint density of $\sum_{i=1}^{n} \psi_1(X_i, \mathbf{t}) \cdots \sum_{i=1}^{n} \psi_p(X_i, \mathbf{t})$ exists and has Fourier transforms that are absolutely integrable both under $f$ and $h_{\mathbf{t}}$. If we let $h_{\mathbf{t},n}$ be the density of $\mathbf{T}_n$ with underlying density $h_{\mathbf{t}}$, then

$$f_n(\mathbf{t}) = c^{-n}(\mathbf{t}) h_{\mathbf{t},n}(\mathbf{t}). \tag{4.23}$$

Proof: Let $\mathbf{Z} = \left( \sum_{k=1}^{n} \psi_1(X_k, \mathbf{t}_0), \cdots, \sum_{k=1}^{n} \psi_p(X_k, \mathbf{t}_0) \right)$ and denote the density of $(\mathbf{Z}, \mathbf{T}_n)$ by $g(\mathbf{z}, \mathbf{t})$ under $f$ and $g_1(\mathbf{z}, \mathbf{t})$ under $h_{\mathbf{t}_0}$. Writing $\mathbf{T}_n = (T_1(\mathbf{x}), \cdots, T_p(\mathbf{x}))$ with $\mathbf{x} = (x_1, \cdots, x_n)$, the moment generating function of $(\mathbf{Z}, \mathbf{T}_n)$ can be written as

$$M(\mathbf{u}, \mathbf{v}) = \int_{\cdots} \int \exp\left\{ \sum_{k=1}^{n} \sum_{j=1}^{p} u_j \psi_j(x_k, \mathbf{t}_0) + \sum_{j=1}^{p} v_j t_j(\mathbf{x}) \right\} \prod_{k=1}^{n} f(x_i) dx_1 \cdots dx_n.$$

Choose $\mathbf{u} = (a_1 + iy_1, a_2 + iy_2, \cdots, a_p + iy_p) = \mathbf{a} + i\mathbf{y}, \mathbf{v} = (iw_1, \cdots, iw_p) = i\mathbf{w}$. Now

$$M(\mathbf{a} + i\mathbf{y}, i\mathbf{w}) = \int_{\cdots} \int \exp\left\{ \sum_{k=1}^{n} \sum_{j=1}^{p} iy_j \psi_j(x_k, \mathbf{t}_0) + \sum_{j=1}^{p} iw_j t_j(\mathbf{x}) \right\}$$

$$\times \exp\left\{ \sum_{k=1}^{n} \sum_{j=1}^{p} a_j \psi_j(x_k, \mathbf{t}_0) \right\} \prod_{k=1}^{n} f(x_k) dx_1 \cdots dx_n$$

$$= c^{-n}(\mathbf{t}_0) \int_{\cdots} \int \exp\left\{ \sum_{k=1}^{n} \sum_{j=1}^{p} iy_j \psi_j(x_k, \mathbf{t}_0) + \sum_{j=1}^{p} iw_j t_j(\mathbf{x}) \right\}$$

$$\prod_{k=1}^{n} h_{\mathbf{t}_0}(x_k) dx_1 \cdots dx_n$$

$$= c^{-n}(\mathbf{t}_0) M_1(i\mathbf{y}, i\mathbf{w}),$$

where $M_1$ is the moment generating function of $(\mathbf{Z}, \mathbf{T}_n)$ under $h_{\mathbf{t}_0}$. Since both $M$ and $M_1$ are absolutely integrable, we can apply the Fourier inversion formula to give

$$g(\mathbf{z}, \mathbf{t}) = \frac{1}{(2\pi i)^{2p}} \int_{\cdots} \int \exp\left\{ -\sum_{j=1}^{p} u_j z_j - \sum_{j=1}^{p} v_j t_j \right\} M(\mathbf{u}, \mathbf{v}) d\mathbf{u} d\mathbf{v},$$

where components of $\mathbf{u}$ and $\mathbf{v}$ are integrated along the path from $c - i\infty$ to $c + i\infty$ for some $c$. Choosing $\mathbf{u} = (\mathbf{a} + i\mathbf{y})$ and $\mathbf{v} = i\mathbf{w}$, we have

$$g(\mathbf{z}, \mathbf{t}) = \frac{1}{(2\pi)^{2p}} \int_{\cdots} \int \exp\left\{ -\sum_{j=1}^{p} (a_j + iy_j) z_j - \sum_{j=1}^{p} iw_j t_j \right\} M(\mathbf{a} + i\mathbf{y}, i\mathbf{w}) d\mathbf{y} d\mathbf{w}$$

$$= \frac{c^{-n}(\mathbf{t}_0)}{(2\pi)^{2p}} \exp\left\{ -\sum_{j=1}^{p} a_j z_j \right\} \int_{\cdots} \int \exp\left\{ \sum_{j=1}^{p} iy_j z_j - \sum_{j=1}^{p} iw_j t_j \right\} M_1(i\mathbf{y}, i\mathbf{w}) d\mathbf{y} d\mathbf{w}$$

$$= c^{-n}(\mathbf{t}_0) \exp\left\{ -\sum_{j=1}^{p} a_j z_j \right\} g_1(\mathbf{z}, \mathbf{t}).$$

Now

$$\mathbf{T}_n(\mathbf{x}) = \mathbf{t}_0 \Longleftrightarrow \sum_{i=1}^{n} \psi_j(x_j, \mathbf{t}_0) = 0, \qquad j = 1, \cdots, p \Longleftrightarrow \mathbf{z} = 0.$$

Hence, $g(\mathbf{z}, \mathbf{t}_0) = c^{-n}(\mathbf{t}_0) g_1(\mathbf{z}, \mathbf{t}_0)$ and from this the result follows. □

We now go through the same steps as in section 4.2 modified for the multivariate case (details given in Field, 1982, p. 675–6). As a result we can write

$$(\mathbf{T}_n - \mathbf{t}_0)_i = \sum\nolimits_j b_{ij} Z_j - \sum_{j,l} \Big\{ \sum_r b_{ir} C_{jl}(r) \Big\} \bar{Z}_j \bar{Z}_l$$
$$+ \sum_{j,l} \sum_r b_{ir} b_{lj} \bar{Z}_j (\bar{Z}_{r_l} - \mu_{r_l}) + 0_p(1/n) \tag{4.24}$$

where $B = -A(t_0)^{-1}$, $A(\mathbf{t}_0) = E_{h_{t_0}} |\frac{\partial \psi}{\partial \mathbf{t}_0}(X, \mathbf{t}_0)|$

$$Z_r = \psi_r(X, \mathbf{t}_0), \quad Z_{rj} = D_j \psi_r(X, \mathbf{t}_0), \quad E(Z_{rj}) = \mu_{rj}$$
$$C_{jl}(r) = \sum_{i_1 i_2} b_{ji_1} b_{li_2} \mu_{ri_1 i_2}.$$

All expected values are with respect to $h_{\mathbf{t}_0}$. Of course the result also holds for $\mathbf{t}_0 = \theta_0$ with density $f$.

Again using the results of James and Mayne (1962) on cumulants, the cumulants behave as in the unvariate case. It then can be shown that the density of $n^{1/2}(\mathbf{T}_n - \mathbf{t}_0)$ under $h_{\mathbf{t}_0}$ at $\mathbf{x}$ is

$$h(\mathbf{x}) = \phi(x)[1 + \sum\nolimits_j d_j D_j \{\phi(\mathbf{x})\}/n^{1/2}$$
$$+ \sum\nolimits_{j,k,l} d_{jkl} D_j D_k D_l \{\phi(\mathbf{x})\}/n^{1/2} + 0(1/n)]$$

where $d_j$ and $d_{jkl}$ are constants determined by the cumulants; $\phi(\mathbf{x})$ is the multivariate p-dimensional normal density with mean 0 and covariance matrix

$$A(\mathbf{t}_0)^{-1} \Sigma(\mathbf{t}_0) (A(\mathbf{t}_0)^{-1})^T \quad \text{and} \quad \Sigma(\mathbf{t}_0) = \{E \psi_r(X, \mathbf{t}_0) \psi_i(X, \mathbf{t}_0)\}_{1 \le r, i \le p}$$

and all expectations are with respect to $h_{\mathbf{t}_0}$. The density of $(\mathbf{T}_n - \mathbf{t}_0)$ under $h_{\mathbf{t}_0}$ at 0 is $h_{\mathbf{t}_0, n}(\mathbf{t}_0) = n^{p/2} h(0)$. From the centering lemma it follows that

$$f_n(\mathbf{t}_0) = (c(\mathbf{t}_0))^{-n} n^{p/2} h(0).$$

Putting the results together gives the multivariate version of Theorem 4.3.

**Theorem 4.5**

If $\mathbf{T}_n$ represents the solution of $\sum_{i=1}^{n} \psi_r(x_i, \mathbf{t}) = 0$, $r = 1, \cdots, p$, and Assumptions A4.1M-A4.5M are satisfied, then an asymptotic expansion for the density of $\mathbf{T}_n$, say $f_n$, is

$$f_n(\mathbf{t}_0) = (n/2\pi)^{p/2} c^{-n}(\mathbf{t}_0) |\det A| |\det \Sigma|^{-1/2} \{1 + 0(1/n)\} \tag{4.25}$$

where $\alpha(\mathbf{t}_0)$ is the solution of

$$\int \psi_r(x,\mathbf{t}_0)\exp\left\{\sum_{j=1}^{p}\alpha_j\psi_j(x,\mathbf{t}_0)\right\}f(x)dx = 0 \quad \text{for} \quad r=1,\cdots,p,$$

$$c^{-1}(\mathbf{t}_0) = \int \exp\left\{\sum_{j=1}^{p}\alpha_j(\mathbf{t}_0)\psi_j(x,\mathbf{t}_0)\right\}f(x)dx,$$

$$A = \left\{E\partial\psi(x,\mathbf{t})/\partial t_r\big|_{\mathbf{t}=\mathbf{t}_0}\right\}_{1\le r,j\le p}, \quad \Sigma = \left\{E\psi_j(x,\mathbf{t}_0)\psi_r(x,\mathbf{t}_0)\right\}_{1\le r,j\le p}$$

and all expectations are with respect to the conjugate density

$$h_{\mathbf{t}_0}(x) = c(\mathbf{t}_0)\exp\left\{\sum_{j=1}^{p}\alpha_j(\mathbf{t}_0)\psi_j(x,\mathbf{t}_0)\right\}f(x).$$

The error term holds uniformly for all $t_0$ in a compact set .

Approximation (4.25) is usually an intermediate step. More often we are interested in tail areas for a marginal distribution of one of the estimates or for some function of the estimates (eg. in tests of hypothesis). In order to do this sort of computations, it is necessary to compute $f_n(\mathbf{t})$ over a grid in $R^p$ and then do a numerical integration to compute the density of $\lambda(\mathbf{T})$ (say). If $p > 3$, it is not computationally feasible to proceed in this fashion and even with $p = 3$, the computational effort required may be large.

As in the univariate case, it would be useful to have (4.25) hold for an arbitrary set so that integrals would still be correct to $0(1/n)$. Although it is probably possible to obtain such results using sophisticated integral approximations, a different approach has been taken in recent work by Tingley and Field (1990) . The details are provided in chapter 6. The techniques there provide a computationally feasible technique for handling higher dimensional problems. The starting point of this approach is the approximation above.

### 4.5.b Calculations

We now turn to the case of location and scale where direct calculations have been carried out with a view of computing percentiles of a studentized version of the location estimate. Given the percentiles, we can then construct approximate confidence intervals for the location parameter. We let $\theta = (\mu,\sigma)$, $f_\theta(x) = f((x-\mu)/\sigma)/\sigma$ and $\psi_i(x,\theta) = \psi_i((x-\mu)/\sigma)$, $i = 1,2$. In particular, we set $\psi_1(x) = \min\{k, \max(-k,x)\}$, $\phi_2(x) = \psi_1^2(x) - \beta$ with $\beta = E_\phi\psi_1^2(x)$.

This corresponds to "Proposal 2" of Huber(1964) and gives translation and scale equivariant estimates.

For $k < \infty$, we have robust M-estimates with a choice of $\beta$ suitable for a model in some neighborhood of the normal. The joint density of $(T_1, T_2)$ was computed giving values in the following table with underlying densities: normal, $t_3$, slash (ratio of normal and uniform on [0, 1]) and Cauchy.

| | | $n=5$ | | | | $n=10$ | | | |
|---|---|---|---|---|---|---|---|---|---|
| $t_1$ | $t_2$ | **normal** | $t_3$ | **slash** | **Cauchy** | **normal** | $t_3$ | **slash** | **Cauchy** |
| 0.00 | .05 | .012856 | .009371 | .007691 | .005387 | .00003 | .000002 | .000001 | .000001 |
| 0.05 | .05 | .007407 | .004776 | .004672 | .002197 | .00001 | 0 | 0 | 0 |
| 1.00 | .05 | .001431 | .000839 | .001082 | .003335 | 0 | 0 | 0 | 0 |
| 1.50 | .05 | .000096 | .000187 | .000107 | .000048 | 0 | 0 | 0 | 0 |
| 2.00 | .05 | 0 | 0 | .000006 | .000009 | 0 | 0 | 0 | 0 |
| 3.00 | .05 | 0 | 0 | 0 | .000001 | 0 | 0 | 0 | 0 |
| 4.00 | .05 | 0 | 0 | 0 | 0 | 0 | 0 | 0 | 0 |
| 0.00 | .50 | .788738 | .536291 | .438449 | .273190 | .504604 | .258916 | .208629 | .078980 |
| 0.50 | .50 | .436232 | .284199 | .286543 | .134394 | .156203 | .074583 | .075231 | .019391 |
| 1.00 | .50 | .074330 | .052233 | .062209 | .023784 | .004716 | .002719 | .003850 | .000663 |
| 1.50 | .50 | .003990 | .005354 | .005707 | .003332 | .000015 | .000032 | .000037 | .000015 |
| 2.00 | .50 | .000070 | .000472 | .000294 | .000564 | 0 | 0 | 0 | 0 |
| 3.00 | .50 | 0 | .000005 | 0 | .000033 | 0 | 0 | 0 | 0 |
| 4.00 | .50 | 0 | 0 | 0 | .000004 | 0 | 0 | 0 | 0 |
| 5.00 | .50 | 0 | 0 | 0 | .000001 | 0 | 0 | 0 | 0 |
| 0.00 | 1.00 | .941091 | .651625 | .693093 | .693093 | 1.882181 | 1.199300 | 1.263642 | .425929 |
| 0.50 | 1.00 | .510072 | .416783 | .427590 | .231780 | .558563 | .488809 | .502669 | .199933 |
| 1.00 | 1.00 | .081463 | .114013 | .107231 | .077477 | .014700 | .036937 | .035675 | .021036 |
| 1.50 | 1.00 | .003873 | .016072 | .013034 | .014750 | .000035 | .000778 | .000614 | .000799 |
| 2.00 | 1.00 | .000056 | .001607 | .000977 | .002492 | 0 | .000009 | .000004 | .000026 |
| 3.00 | 1.00 | 0 | .000017 | – | .000127 | 0 | 0 | 0 | 0 |
| 4.00 | 1.00 | 0 | 0 | .000005 | .000014 | 0 | 0 | 0 | 0 |
| 5.00 | 1.00 | 0 | 0 | 0 | .000002 | 0 | 0 | 0 | 0 |
| 0.00 | 2.00 | .017740 | .092733 | .076215 | .110021 | .001786 | .098294 | .081975 | .218988 |
| 0.50 | 2.00 | .009514 | .080232 | .967794 | .102566 | .000516 | .071258 | .064659 | .180062 |
| 1.00 | 2.00 | .001468 | .049090 | .043112 | .078223 | .000012 | .024586 | .025895 | .090319 |
| 1.50 | 2.00 | .000065 | .018907 | .016886 | .041991 | 0 | .003333 | .003854 | .021953 |
| 2.00 | 2.00 | .000001 | .004405 | .003957 | .014694 | 0 | .000172 | .000198 | .002430 |
| 3.00 | 2.00 | 0 | .000080 | .000074 | .000900 | 0 | 0 | .000004 | .000010 |
| 4.00 | 2.00 | 0 | .000002 | .000001 | .000069 | 0 | 0 | 0 | 0 |
| 5.00 | 2.00 | 0 | 0 | 0 | .000010 | 0 | 0 | 0 | 0 |
| 0.00 | 3.00 | .000003 | .008352 | .0078537 | .033976 | 0 | .002132 | .002132 | .055246 |
| 0.50 | 3.00 | .000002 | .008156 | .008281 | .033681 | 0 | .001962 | .003645 | .052255 |
| 1.00 | 3.00 | 0 | .007339 | .009137 | .032197 | 0 | .001446 | .003787 | .042583 |
| 1.50 | 3.00 | 0 | .005824 | .008654 | .027821 | 0 | .000701 | .002738 | .026525 |
| 2.00 | 3.00 | 0 | .002925 | .005405 | .019309 | 0 | .000172 | .00086 | .010439 |
| 3.00 | 3.00 | 0 | .000241 | .000489 | .003687 | 0 | .000001 | .000005 | .000303 |
| 4.00 | 3.00 | 0 | .000007 | .000016 | .000359 | 0 | 0 | 0 | .000003 |
| 5.00 | 3.00 | 0 | 0 | .000001 | .000037 | 0 | 0 | 0 | 0 |

| $t_1$ | $t_2$ | normal ($n=5$) | $t_3$ ($n=5$) | slash ($n=5$) | Cauchy ($n=5$) | normal ($n=10$) | $t_3$ ($n=10$) | slash ($n=10$) | Cauchy ($n=10$) |
|---|---|---|---|---|---|---|---|---|---|
| 0.00 | 4.00 | 0 | .000985 | .001927 | .012927 | .012603 | 0 | .000063 | .015383 |
| 0.50 | 4.00 | 0 | .001001 | .001927 | .012679 | 0 | .000063 | .000475 | .015168 |
| 1.00 | 4.00 | 0 | .111034 | .002205 | .012830 | 0 | .000055 | .000647 | .012404 |
| 2.00 | 4.00 | 0 | .000913 | .002960 | .011952 | 0 | .000032 | .000604 | .008956 |
| 3.00 | 4.00 | 0 | .000274 | .001183 | .006007 | 0 | .000002 | .000058 | .001541 |
| 4.00 | 4.00 | 0 | .000021 | .000010 | .001190 | 0 | 0 | 0 | .000052 |
| 5.00 | 4.00 | 0 | .000001 | .000005 | .000160 | 0 | 0 | 0 | .000001 |
| 0.00 | 5.00 | 0 | .000162 | .000645 | .005495 | 0 | .000003 | .000101 | .005064 |
| 1.00 | 5.00 | 0 | .000179 | .000743 | .005692 | 0 | .000003 | .000122 | .005040 |
| 3.00 | 5.00 | 0 | .000168 | .001259 | .005229 | 0 | .000002 | .000125 | .002213 |
| 5.00 | 5.00 | 0 | .000005 | .000028 | .000460 | 0 | 0 | 0 | .000012 |
| 10.00 | 5.00 | 0 | 0 | 0 | 0 | 0 | 0 | 0 | 0 |
| 0.00 | 10.00 | 0 | 0 | .000031 | .000346 | 0 | 0 | .000001 | .000109 |
| 1.00 | 10.00 | 0 | 0 | .000032 | .000355 | 0 | 0 | .000001 | .000113 |
| 3.00 | 10.00 | 0 | 0 | .000048 | .000431 | 0 | 0 | .000001 | .000137 |
| 5.00 | 10.00 | 0 | 0 | .000114 | .000597 | 0 | 0 | .000003 | .000151 |
| 10.00 | 10.00 | 0 | 0 | .000001 | .000018 | 0 | 0 | 0 | 0 |
| 0.00 | 15.00 | 0 | 0 | .000006 | .000065 | 0 | 0 | 0 | .000010 |
| 1.00 | 15.00 | 0 | 0 | .000006 | .000066 | 0 | 0 | 0 | .000010 |
| 3.00 | 15.00 | 0 | 0 | .000007 | .000074 | 0 | 0 | 0 | .000012 |
| 5.00 | 15.00 | 0 | 0 | .000010 | .000094 | 0 | 0 | 0 | .000016 |
| 10.00 | 15.00 | 0 | 0 | .000017 | .000116 | 0 | 0 | 0 | .000007 |
| 0.00 | 20.00 | 0 | 0 | 0 | .000019 | 0 | 0 | 0 | .000002 |
| 1.00 | 20.00 | 0 | 0 | 0 | .000020 | 0 | 0 | 0 | .000002 |
| 3.00 | 20.00 | 0 | 0 | 0 | .000022 | 0 | 0 | 0 | .000002 |
| 5.00 | 20.00 | 0 | 0 | 0 | .000025 | 0 | 0 | 0 | .000003 |
| 10.00 | 20.00 | 0 | 0 | 0 | .000056 | 0 | 0 | 0 | .000006 |
| 15.00 | 20.00 | 0 | 0 | 0 | .000016 | 0 | 0 | 0 | 0 |

**Exhibit 4.12**
Approximation to the joint density of robust estimates
of location and scale using Huber's proposal 2 with $k = 1.5$.

To check the accuracy of the results is difficult since there is no obvious method for computing the exact joint density. The approximation can be checked for the marginal density of $T_1$ using Monte Carlo results from the Princeton Robustness Study, reported in part by Andrews et al (1971). The complete results have been provided most kindly by F. Hampel. The values of the pseudovariances and $n$ times the variance are reported both for the approximation (obtained by integrating numerically the joint density) and the Monte Carlo results in Exhibit 4.13. The pseudovariance is defined as $n(t_{1,1-\alpha}/z_{1-\alpha})^2$ where $t_{1,1-\alpha}$ and $z_{1-\alpha}$ represent the $(1-\alpha)$ quantile of the distribution of the estimator $T_1$ and a standard normal variate respectively.

| | $n=5$ | | $n=10$ | | $n=20$ | | $n=40$ | |
|---|---|---|---|---|---|---|---|---|
| **Normal** | **A** | **MC** | **A** | **MC** | **A** | **MC** | **A** | **MC** |
| Pseudo-variances: | | | | | | | | |
| 25% | 1.0345 | 1.0412 | 1.0357 | 1.0312 | 1.0368 | 1.036 | 1.0380 | 1.0392 |
| 10% | 1.0353 | 1.0411 | 1.0360 | 1.0303 | 1.0366 | 1.0356 | 1.0373 | 1.0380 |
| 2.5% | 1.0367 | 1.0432 | 1.0366 | 1.0306 | 1.0368 | 1.0357 | 1.0372 | 1.0380 |
| 1% | 1.0449 | 1.0462 | 1.0371 | 1.0310 | 1.0370 | 1.0359 | 1.0372 | 1.0382 |
| .5% | 1.0385 | 1.0462 | 1.0374 | 1.0313 | 1.0372 | 1.0361 | 1.0372 | 1.0383 |
| .1% | 1.0398 | 1.0496 | 1.0384 | 1.0321 | 1.0376 | 1.0366 | 1.0373 | 1.0386 |
| n×var | 1.0360 | 1.0427 | 1.0364 | 1.0308 | 1.0369 | 1.0360 | 1.0375 | 1.0384 |
| **Slash** | | | | | | | | |
| Pseudo-variances: | | | | | | | | |
| 25% | 1.7610 | 1.8597 | 1.6856 | 1.6599 | 1.6457 | 1.6559 | 1.6284 | 1.5953 |
| 10% | 1.9548 | 2.0715 | 1.7624 | 1.7265 | 1.6758 | 1.6902 | 1.6392 | 1.6076 |
| 2.5% | 2.8046 | 2.8036 | 1.9878 | 1.8941 | 1.7529 | 1.7679 | 1.6711 | 1.6341 |
| 1% | 4.5851 | 4.2350 | 2.2613 | 2.0682 | 1.8283 | 1.8234 | 1.6962 | 1.6535 |
| .5% | 7.3134 | 6.7839 | 2.5974 | 2.2829 | 1.8880 | 1.8876 | 1.7166 | 1.6686 |
| .1% | 18.4117 | 19.7112 | 4.3116 | 4.5852 | 2.1149 | 2.0288 | 1.7755 | 1.7040 |
| n×var | 3.549 | 3.8752 | 2.0776 | 3.5681 | 1.7419 | 1.7986 | 1.6629 | 1.6246 |
| **Cauchy** | | | | | | | | |
| Pseudo-variances: | | | | | | | | |
| 25% | 4.607 | 3.75 | 4.590 | 4.5731 | 4.907 | 4.648 | 4.4852 | 4.0 |
| | | | | 4.7400* | | | | |
| 10% | 7.256 | 5.4060 | 5.834 | 5.8120 | 5.094 | 4.8625 | 4.7554 | 4.2781 |
| | | | | 6.1554* | | | | |
| 2.5% | 17.405 | 11.590 | 9.392 | 9.2350 | 6.429 | 5.8338 | 5.3361 | 6.6673 |
| | | | | 9.3463* | | | | |
| 1% | 30.747 | 19.2729 | 13.434 | 14.6001 | 7.634 | 6.8401 | 5.8013 | 4.9629 |
| | | | | 12.1464* | | | | |
| .5% | 44.252 | 26.6878 | 17.897 | 21.2734 | 8.752 | 7.7168 | 6.1902 | 5.2147 |
| | | | | 14.9575* | | | | |
| .1% | 72.847 | 43.9043 | 35.365 | 47.2187 | 12.45 | 10.1046 | 7.2629 | 6.1221 |
| | | | | 39.2633* | | | | |
| n×var | 16.525 | 10.9373 | 9.592 | 10.2658 | 6.172 | 5.6630 | 5.161 | 4.5469 |
| $t_3$ | | | | *replication | | | | |
| Pseudo-variances: | | | | | | | | |
| 25% | 1.6275 | | 1.6569 | | 1.6556 | 1.655 | 1.6522 | |
| 10% | 1.7670 | | 1.7226 | | 1.6870 | 1.6858 | 1.6671 | |
| 2.5% | 2.0690 | | 1.8442 | | 1.7488 | 1.7348 | 1.6968 | |
| 1% | 2.3447 | | 1.9611 | | 1.7948 | 1.7652 | 1.7158 | |
| 5% | 2.5572 | | 2.0478 | | 1.8323 | 1.7873 | 1.7360 | |
| 1% | 3.2904 | | 2.2680 | | 1.9245 | 1.8370 | 1.7789 | |
| n×var | 1.9953 | | 1.8097 | | 1.7273 | 1.7132 | 1.6870 | |

**Exhibit 4.13**

Pseudovariances and asymptotic variances of $T_1$ as computed by approximation ($A$) and Monte Carlo ($MC$)

In order to determine whether the difference between the Monte Carlo result and the approximation are to within the sampling errors in the Monte Carlo experiment, we can look at two bits of evidence. In Exhibit 5.13, Andrews et al. (1971) give differences between exact and Monte Carlo results for the pseudo variance of the median, with $n = 5$. With the normal, the differences are in the range of .02 to .03. The differences in Exhibit 4.13 between the approximate and Monte Carlo pseudovariances for the normal, $n = 5$, all are less than .01 indicating differences are well within the errors inherent in the Monte Carlo results. For the Cauchy, the differences observed in Exhibit 4.13 are larger by a factor of up to 10 than those reported for the median (cf. Table 5, Field (1982)). To shed light on whether we can place any faith in the asymptotic results for the Cauchy, it is worth looking at the Monte Carlo results for $n = 10$. For this situation, there were two simulations carried out in the Princeton study and the replication gives an indication of the Monte Carlo errors. From Exhibit 4.13, with the Cauchy and $n = 10$, the asymptotic results lie between the two Monte Carlo replications except for 0.1% and $n\times$ variance. This gives a strong indication of the reliability of the asymptotic results for $n = 10$. Until the exact marginal densities are computed in some fashion, or until additional Monte Carlo studies are done, further comparisons are difficult. To see that the large discrepancies for the Cauchy at $n = 5$ may be due to Monte Carlo variation, it is instructive to look at Exhibit 4.14.

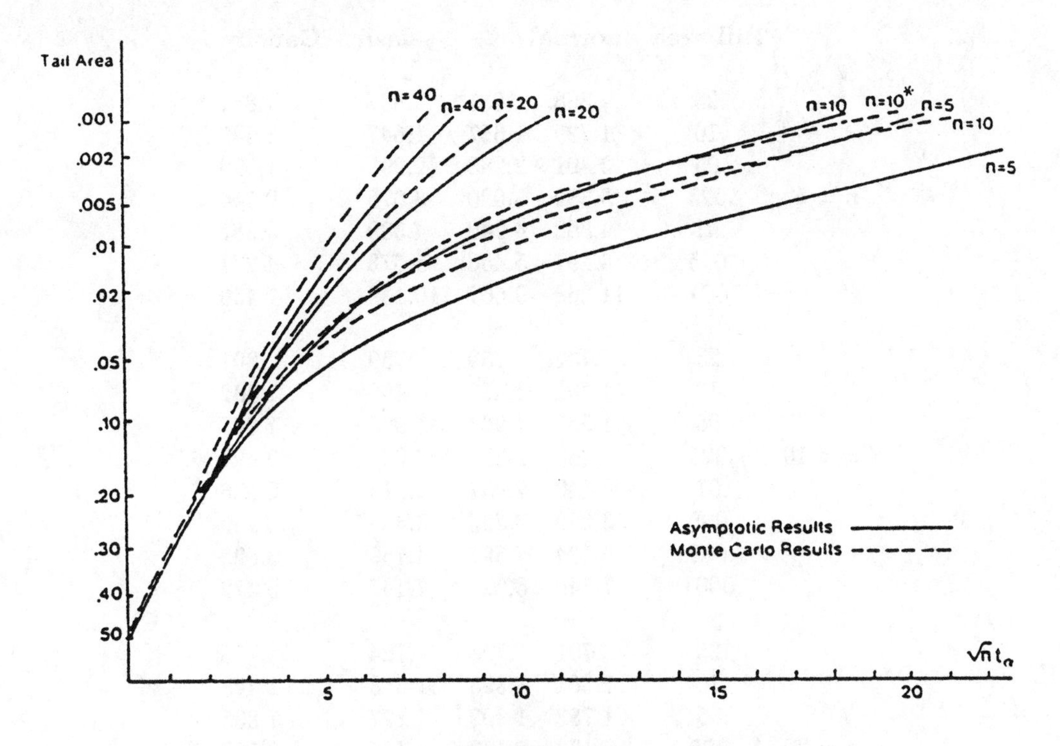


**Exhibit 4.14**

Plot of percentiles of $\sqrt{n}T_1$ on normal probability paper for Huber's Proposal 2, $k = 1.5$.

From the graph, we note that the Monte Carlo results for $n = 5$ do not follow the pattern exhibited by the other values of $n$. In particular, it appears that the extreme percentiles for the Monte Carlo with $n = 5$ are not large enough. This would lead to the large differences observed in Exhibit 4.13. While this is not a proof that the asymptotic results are accurate, it suggests that the precision of the asymptotic results may be very good even in the extreme case of the Cauchy with $n = 5$.

Given the approximation to the joint density, it is possible to examine in detail interesting characteristics of the robust estimates of location and scale. We illustrate the potential with two different computations.

To continue with this example, we consider the percentiles of a "studentized" version of $T_1$. Since the asymptotic variance $T_1$ is

$$\sigma^2 E_f \psi^2(X) / \left\{ E_f \psi'(X) \right\}^2,$$

an appropriate "studentized" version of $T_1$ would be $n^{1/2} T_1 / (T_2 \gamma)$ with $\gamma = E_\phi \psi_1^2(X) / \left\{ E_\phi \psi_1' \right\}^2$. This assumes that the estimate has been chosen as though the underlying density is normal. This was implicit in the definition of $\psi_2$ at the beginning of the example. In practice, we could replace $\gamma$ by its estimated form where $\Phi$ is replaced by the empirical distribution. However the problem of working out the percentiles of this more complicated expression introduces some computational difficulties.

The percentiles have been evaluated by numerical integration of the joint density of $T_1$ and $T_2$ over the appropriate region of the plane. The results are tabulated in Exhibit 4.15.

| | Tail area | normal | $t_3$ | slash | Cauchy |
|---|---|---|---|---|---|
| | .25 | .808 | .831 | .837 | .871 |
| | .10 | 1.729 | 1.657 | 1.647 | 1.547 |
| | .05 | 2.491 | 2.288 | 2.297 | 1.999 |
| $n = 5$ | .025 | 3.382 | 3.020 | 3.075 | 2.514 |
| | .01 | 4.860 | 4.249 | 4.389 | 3.387 |
| | .005 | 6.297 | 5.456 | 5.673 | 4.271 |
| | .001 | 11.269 | 9.667 | 10.139 | 7.450 |
| | .25 | .732 | .759 | .759 | .807 |
| | .10 | 1.461 | 1.467 | 1.468 | 1.482 |
| | .05 | 1.965 | 1.923 | 1.925 | 1.863 |
| $n = 10$ | .025 | 2.460 | 2.358 | 2.369 | 2.197 |
| | .01 | 3.143 | 2.937 | 2.975 | 2.626 |
| | .005 | 3.689 | 3.393 | 3.464 | 2.959 |
| | .001 | 5.124 | 4.597 | 4.759 | 3.828 |
| | .0001 | 7.749 | 6.823 | 7.147 | 5.472 |
| | .25 | .701 | .728 | .724 | .775 |
| | .10 | 1.361 | 1.393 | 1.388 | 1.448 |
| | .05 | 1.783 | 1.802 | 1.797 | 1.835 |
| $n = 20$ | .025 | 2.173 | 2.168 | 2.166 | 2.162 |
| | .01 | 2.665 | 2.613 | 2.620 | 2.536 |
| | .005 | 3.027 | 2.937 | 2.953 | 2.795 |
| | .001 | 3.866 | 3.664 | 3.714 | 3.395 |
| | .0001 | 5.118 | 4.726 | 4.845 | 4.139 |

**Exhibit 4.15a**

Percentiles of $n^{1/2} T_1 / \gamma T_2$ using Huber's Proposal 2 with $k = 1.5$

| | Tail area | normal | $t_3$ | slash | Cauchy |
|---|---|---|---|---|---|
| | .25 | .686 | .713 | .707 | .759 |
| | .10 | 1.318 | 1.359 | 1.250 | 1.430 |
| | .05 | 1.709 | 1.750 | 1.741 | 1.824 |
| $n = 40$ | .025 | 2.060 | 2.095 | 2.086 | 2.156 |
| | .01 | 1.477 | 2.500 | 2.492 | 2.540 |
| | .005 | 2.781 | 2.785 | 2.781 | 2.794 |
| | .001 | 3.433 | 3.388 | 3.394 | 3.317 |
| | .0001 | 4.307 | 4.174 | 4.205 | 3.959 |
| | .25 | .679 | .705 | .698 | .750 |
| | .10 | 1.296 | 1.341 | 1.329 | 1.419 |
| | .05 | 1.668 | 1.722 | 1.708 | 1.817 |
| $n = 100$ | .025 | 1.993 | 2.055 | 2.039 | 2.158 |
| | .01 | 2.383 | 2.445 | 2.428 | 2.550 |
| | .005 | 2.649 | 2.710 | 1.693 | 2.917 |
| | .001 | 3.213 | 3.266 | 3.250 | 3.355 |
| | .0001 | 3.922 | 3.953 | 3.941 | 4.005 |

**Exhibit 4.15b**

Percentiles of $n^{1/2}T_1/\gamma T_2$ using Huber's Proposal 2 with $k = 1.5$

The first thing to check in Exhibit 4.15 is the agreement of the percentiles under the normal with the percentiles of a t-density. There is a good, but not perfect, agreement with the t-density for degrees of freedom about $0.6n$. This seems to hold over the whole range of $n$ values from 5 to 100. This result confirms some speculation that the "studentized" ratios behave like a t-density with reduced degrees of freedom, but the reduction may be larger than expected.

The important question of the stability of the percentiles as the underlying density varies can be examined using these results. As is to be expected, the largest variation occurs with small n and a Cauchy density. For $n = 5$, if we computed a 99% confidence interval, based on the normal figures, the interval would be 1.43 times longer than the correct interval for a Cauchy density while a 99.99% confidence interval would be 1.51 times longer than necessary. These results, as they are, are an order of magnitude improvement over results using a classical t-interval.

As a second computation we consider the question of the degree of dependence between $T_1$ and $T_2$. For the normal with estimates $\bar{x}$ and $s$, we have independence and it is interesting to compare the behavior of $T_1$ and $T_2$ with this. There is no standard measure of dependence between two random variables. Renyi (1959) has proposed several measures which satisfy most of the properties he feels are natural. We compute two of these measures for the joint distribution of $(T_1, T_2)$. The first of these $\gamma_n$ is a normalized version of the mean square contingency,

$$C_n = \left(\int\int (k(x,y) - 1)^{-2} dP_{n,1}(x) dP_{n,2}(y)\right)^{1/2}$$

with $k(x,y) = p_n(x,y)/p_{n,1}(x)p_{n,2}(y)$, given as $\Gamma_n = C_n/(1 + C_n^2)^{1/2}$. The second measure $L_n$ is based on information theoretical considerations and can be written as

$$L_n(T_1, T_2) = (1 - \exp(-2I(T_1, T_2))^{1/2}$$

where $I(T_1,T_2) = \int\int k(x,y)\log k(x,y)dP_{n,1}(x)dP_{n,2}(y)$ is the amount of information $T_1$ contains about $T_2$.

The calculations have been done for $(T_1,T_2)$ giving the following results in Exhibit 4.16.

| | normal | | $t_3$ | | Cauchy | |
|---|---|---|---|---|---|---|
| n | $\Gamma_n$ | $L_n$ | $\Gamma_n$ | $L_n$ | $\Gamma_n$ | $L_n$ |
| 5 | .021 | .017 | .101 | .090 | .351 | .326 |
| 10 | .015 | .016 | .071 | .067 | .283 | .271 |
| 15 | .010 | .010 | .056 | .054 | .241 | .233 |
| 20 | .008 | .008 | .048 | .047 | .212 | .207 |
| 25 | .007 | .007 | .043 | .042 | .192 | .187 |
| 30 | .007 | .007 | .039 | .038 | .176 | .171 |
| 35 | .006 | .006 | .036 | .035 | .163 | .158 |
| 40 | .006 | .006 | .033 | .032 | .152 | .147 |
| 45 | .005 | .005 | .031 | .031 | .143 | .137 |
| 50 | .005 | .005 | .030 | .029 | .135 | .129 |

**Exhibit 4.16**
Dependence measures for $(T_1,T_2)$

There are several interesting features of these results including the considerable variation in the dependence structure for different underlying densities. It is perhaps surprising to note such differences for rather similar underlying densities. Looking at the results as $n$ increases, the dependence measures seem to be approaching 0 at a rather slow rate. For the case of the normal, it appears that $\Gamma_n$ is decreasing at a rate of $1/n$ and in fact the relationship $\Gamma_n = 1/(10n)$ gives a good fit to the data.

The purpose in this example has not been to carry out an extensive study of all interesting properties of the robust location/scale but rather to illustrate the potential of the approximation for examining these types of questions.

It is worth noting that if we set $k = \infty$, we obtain the classical estimates of location and scale. If the underlying density is $N(0,1)$, then the equation of Theorem 4.5 can be solved explicitly giving $\alpha_1(\mathbf{t}) = t_1t_2$ and $\alpha_2(\mathbf{t}) = (t_2^2-1)/2$. The conjugate density $h_{\mathbf{t}}(y)$ is normal with mean $t_1$ and variance $t_2^2$. It is easy to show that the approximating formula (4.25) becomes

$$f_n(t_1,t_2) = (n/2\pi)t_2^{n-2}\exp(-nt_2^2/2 - nt_1^2/2 + n^{1/2})2^{1/2}.$$

This agrees with the exact formula except for the constant terms which are in the ratio $n^{n/2-1}\pi^{1/2}2^{3/2-n/2}e^{-n/2}/\Gamma((n-1)/2)$. For $n = 9$, this ratio equals .897 so that the error from the constant term is relatively large, emphasizing the need for a numerical rescaling of the approximation.

### 4.5.c Regression

We now consider modifying the approximation developed in the previous section to the regression case. Let $y_i = \eta_i(\theta) + u_i$ be $n$ iid random variables with scale parameter $\sigma$. The estimates $\mathbf{T} = (T_1,\cdots,T_{p+1})$ of $(\theta,\sigma)$ are the values which minimize (Huber, 1981, Ch.14)

$Q(\theta,\sigma) = \sum_{i=1}^{n} \rho((y_i - \eta_i(\theta))/\sigma)\sigma + a\sigma;\ \sigma \geq 0$ or equivalently solve

$$\sum_{i=1}^{n} \psi((y_i - \eta_i(\theta))/\sigma)\partial\eta_i/\partial\theta_j = 0, \quad j = 1,\cdots,p \tag{4.26}$$

$$\sum_{i=1}^{n} \chi((y_i - \eta_i(\theta))/\sigma) - a = 0$$

where $\psi = \rho'$ and $\chi(x) = x\psi(x) - \rho(x)$. Since the observations $y_i$ are not identically distributed, the results of the previous section cannot be applied directly but must be modified. Note that $\rho(x) = x^2/2$ gives the standard least squares estimates. To begin the modifications required, let $\mathbf{t}$ be the point at which the density is to be evaluated, $\Delta_i = y_i - \eta_i(\mathbf{t})$, $f_i$ the density of $y_i$, where $f_i$ depends on the underlying value of $(\theta,\sigma)$, say $(\theta_{10},\cdots,\theta_{p0},\sigma_0)$, $z_j$, $j = 1,\cdots,p+1$, the left hand side of the equations in (4.26). To proceed with the centering result (cf. 4.23), let the conjugate density for the $i^{th}$ observation be

$$h_{\mathbf{t}}^i(y) = c^i(\mathbf{t})f_i(y)\exp\left\{\sum_{j=1}^{p} \alpha_j\psi(\Delta_i/t_{p+1})\partial\eta_i/\partial t_j + \alpha_{p+1}\chi(\Delta_i/t_{p+1})\right\}$$

where $c^i(\mathbf{t})$ is the appropriate normalizing constant so that $\int h_{\mathbf{t}}^i(y)dy = 1$. Then it follows that $\phi(\alpha + i\mathbf{y}, i\mathbf{v}) = \prod_{i=1}^{n} c^i(\mathbf{t})M_1(i\mathbf{y}, i\mathbf{v})$ where $M(M_1)$ is the moment generating function of $(z_1,\cdots,z_{p+1},T_1,\cdots,T_{p+1})$ under density $f_i(h_{\mathbf{t}}^i)$, $i = 1,\cdots,n$. From this it follows that $f_n(\mathbf{t}) = \left(\prod_{i=1}^{n} c^i(\mathbf{t})\right)^{-1} h_{\mathbf{t},n}(\mathbf{t})$ where $h_{\mathbf{t},n}$ is the density under the conjugate density $(g_{\mathbf{t}}^1,\cdots,g_{\mathbf{t}}^n)$. The vector $\alpha$ solves the following set of $p+1$ equations in $p+1$ unknowns.

$$\sum_{i=1}^{n} \int \psi((y_i - \eta_i(\mathbf{t}))/t_{p+1})\partial\eta_i(\mathbf{t})/\partial t_j h_{\mathbf{t}}^i(y_i)dy_i = 0, \quad j = 1,\cdots,p$$

$$\sum_{i=1}^{n} \int \left[\chi((y_i - \eta_i(\mathbf{t}))/t_{p+1}) - a\right] h_{\mathbf{t}}^i(y_i)dy_i = 0 \tag{4.27}$$

The arguments expressing $\mathbf{T}_n$ as a mean and obtaining the multivariate expansion given in the previous section go through with minor notational changes. This leads to the following approximating density for $(T_1,\cdots,T_p,T_{p+1})$ where $T_1,\cdots,T_p$ are the estimates of $\theta$ and $T_{p+1}$ estimates $\sigma$:

$$f_n(\mathbf{t}_0) = (n/2\pi)^{p/2}\left(\prod_{i=1}^{n} c_{\mathbf{t}_0}^i\right)^{-1} |\det A||\det\Sigma|^{-1/2}$$

where

$$A = \left\{E\frac{\partial}{\partial t_\ell}Z_r\Big|_{\mathbf{t}=\mathbf{t}_0}\right\}_{1\leq\ell,r\leq p+1} \tag{4.28}$$

and $\sum = \{EZ_\ell Z_r\}_{1\leq\ell,r\leq p+1}$ with the expectations $E\frac{\partial}{\partial t_\ell}z_r$ to be interpreted as $E\frac{\partial}{\partial t_\ell}\sum_{i=1}^{n} E_{h_{\mathbf{t}}^i}\left[\psi((y_i - f_i(\mathbf{t}))/t_{p+1})\partial f_i/\partial t_r\right]$ if $r \leq p$ and with $\chi$ in the square brackets for $r = p+1$. Similar interpretations hold for $EZ_\ell Z_r$.

As a special case, assume $\rho(x) = x^2/2$, $a = n-2$, $\eta_i(\theta) = \theta_1 + \theta_2(x_i - \bar{x})$, $u_i$'s are independent $N(0,\sigma^2)$. The solution of the equations (4.26) yields the least squares estimates for straight line regression. Equations (4.27) become

$$\sum_{i=1}^{n} \int (y_i - t_1 - t_2(x_i - \bar{x}))h_t^i(y_i)dy_i/t_3 = 0$$

$$\sum_{i=1}^{n} \int (y_i - t_1 - t_2(x_i - \bar{x}))(x_i - \bar{x})h_t^i(y_i)dy_i/t_3 = 0$$

$$\sum_{i=1}^{n} \int \left[(y_i - t_1 - t_2(x_i - \bar{x}))^2/t_3^2 - (n-2)\right] h_t^i(y_i)dy_i = 0.$$

It can be seen that these equations will be satisfied if $h_t^i(y)$ is $N(t_1+t_2(x_i-x),\,(n-2)t_3^2/n)$. By choosing $\alpha_1(\mathbf{t}) = (t_1-\theta_1)t_3/\sigma^2$, $\alpha_2(\mathbf{t}) = (t_2-\theta_2)t_3/\sigma^2$, $\alpha_3(\mathbf{t}) = (t_3^2/2\sigma^2 - n/2(n-2)$, $h_t^i(y)$ is $N(t_1+t_2(x_i-\bar{x}),\ (n-2)t_3^2/n)$ when $f_i(y)$ is $N(\theta_1+\theta_2(x_i-x),\sigma^2)$. Evaluating $c^i(\mathbf{t})$, we obtain

$$\prod_{i=1}^{n} c^i(\mathbf{t}) = ((n-2/n)^{n-2}(t_3/\sigma)^n$$

$$\exp\left\{-n(t_1-\theta_1)^2/2\sigma^2 - (t_2-\theta_2)^2 \sum_{i=1}^{n}(x_i-\bar{x})^2/2\sigma^2 - (n-2)t_3^2/2\sigma_0^2 + n^2\right\}.$$

It can be shown that $\det A = (n-2)\sum_{i=1}^{n}(x_i-\bar{x})^2/2t_3^2$, $(\det\Sigma)^{1/2} \propto (\sum_{i=1}^{n}(x_i-\bar{x})^2)^{1/2}$. The approximating density (4.28) evaluated at $\mathbf{t} = (t_1,t_2,t_3)$ with underlying observations $Y_1,\cdots,Y_n$ where $Y_i$ is $N(\theta_1+\theta_2(x_i-\bar{x}),\ \sigma^2)$ is:

$$f_n(\mathbf{t}) = (t_3/\sigma)^n$$

$$\exp\{-n(t_1-\theta_1)^2/2\sigma^2 - (t_2-\theta_2)^2\sum_{i=1}^{n}(x_i-\bar{x})^2/2\sigma^2 - (n-2)t_3^2/2\sigma^2\}$$

$$\left(\sum_{i=1}^{n}(x_i-\bar{x})^2\right)^{1/2}/t_3^2$$

up to a constant of integration. This agrees with the exact density up to the constant of integration.

For other choices of $\rho$, it is not possible to find explicit solutions. Even for the case of linear regression, we have to evaluate $f_n$ over a three-dimensional grid. In many situations we are interested in the marginal density of one of the parameters. It's not usually feasible to evaluate $f_n(\mathbf{t})$ over a grid and then numerically integrate to obtain the marginal density. In section 6.3 we discuss a procedure which provides a one-dimensional technique to construct confidence intervals for a function of the parameters in multiparameter problem.

Spady (1987) computes the saddlepoint approximation to the density of a symmetrically trimmed least squares estimator for the censored regression model with an intercept and one regressor.

# 5. RELATED TECHNIQUES

## 5.1 INTRODUCTION

In this chapter, a number of related techniques will be presented emphasizing their relationship to small sample asymptotics. Section 2 looks at an approach developed by Frank Hampel which has a number of desirable properties.

Next the relationship of small sample asymptotics to saddlepoint and large deviations is presented. We then turn to the work of Durbin and Barndorff-Nielsen and attempt to relate their work in the case of sufficiency and/or exponential families to the techniques of small sample asymptotics. To conclude the chapter, computations are done in the case of logistic regression to contrast the various approaches.

## 5.2. HAMPEL'S TECHNIQUE

In the paper, Hampel (1973), many of the motivating ideas for small sample asymptotics are laid down. Both authors were introduced to the topic via the paper and it is important to acknowledge its influence. Although the results turn out to be closely related to saddlepoint results, they were developed independently of the saddlepoint work of Daniels (1954). The approach proposed by Hampel is very interesting and probably has yet to be fully exploited. Our purpose here is to present the ideas and suggest some possible future directions. The initial development follows Hampel (1973) very closely, especially p. 111, 112.

Hampel's approach differs in several ways from typical classical approaches. The first is that the density of the estimate, rather than a standardized version of it, is approximated. A second feature is that we use low-order expansion in each point separately and then integrate the results rather than use a high-order expansion around a single point. It is this feature which really distinguishes small sample asymptotics (and saddlepoint techniques) from classical asymptotic expansions. The local accuracy from the first one or two terms is effectively transferred to a selected grid of points yielding the same accuracy globally. It is the availability of cheap computing which makes feasible this use of local techniques. A fairly simple approximation requiring non-trivial computation is carried out at a number of grid points. This is of course exactly the type of problem which is ideally suited to computer computations.

The third difference concerns the question of what to expand. Hampel argues effectively that the most natural and simple quantity to study is the derivative of the logarithm of the density, namely $f_n'/f_n$. There are at least four reasons why this seems reasonable.

(i) The form of the expansion of $f_n'/f_n$ is such that the first term is proportional to $n$ and the first two terms are linear in $n$. This contrasts with more complicated relationships coming from $f_n$ or the cumulative.

(ii) Since our expansions are local in nature, it makes sense to focus on a feature of a distribution which is not affected by shifts or addition or deletion of mass elsewhere. Neither $f_n$ or the cumulative satisfy these properties. $f_n'/f_n$ is the first and simplest quantity with these local properties.

(iii) We can view the normal distribution as playing a very special and basic role in probability, in many ways analogous to the role of the straight line in geometry. For the normal, it is $f'/f$ which has a particularly simple form, namely a linear function of $x$. By expanding $f_n'/f_n$ locally, we are, in a sense, linearizing a function locally.

(iv) An expansion of $f_n'/f_n$ will not give the constant of integration for $f_n$ but forces us to determine it numerically. As has been noted in Remark 3.2, in approximating the density of mean, the order of the error is improved from $n^{-1}$ to $n^{-3/2}$ by renormalization. Using the $f_n'/f_n$ scale emphasizes the renormalization in a natural way.

To contrast this approach to the techniques developed so far, namely, it is useful to consider a specific problem. Consider the situation of approximating $f_n'/f_n$ for the mean of $n$ independent observations. The presentation is similar to that found in Field (1985). The development for the more important case of M-estimates of location is given in Field and Hampel (1982).

Assume that $f_n'/f_n$ is to be approximated at a point $t$. The conjugate density is $h_t(x) = c(t)\exp\{\alpha(t)(x-t)\}f(x)$ and $\alpha(t)$ is the solution of

$$\int (x-t)h_t(x)dx = 0 \quad \text{or} \quad \int (x-t)\exp\{\alpha(t)(x-t)\}f(x)dx = 0.$$

In order to guarantee the existence of $\alpha(t)$ and its derivatives up to order 4, assume that $\int x^r e^{\alpha x} f(x)dx$ exists for $r$ up to 5.

Now we obtain a centering results (cf 4.4, 4.23) as follows:

$$\begin{aligned} f_n(t) &= n \int_{\dots}\int f(nt - \sum_1^{n-1} x_i)\prod_1^{n-1} f(x_i)d\mathbf{x} \\ &= nc^{-n}(t)\int_{\dots}\int h_t(nt - \sum_1^{n-1} x_i)\prod_1^{n-1} h_t(x_i)d\mathbf{x} \\ &= c^{-n}(t)h_{t,n}(t) \end{aligned}$$

where $h_{t,n}(t)$ is the density of $\bar{X}$ with underlying density $h_t$. Now we use a normal approximation to $\bar{X}$ under $h_t$. Recall $E_{h_t}\bar{X} = t$ and $\text{var}_{h_t}\bar{X} = \int (x-t)^2 h_t(x)dx/n \equiv \sigma^2(t)/n$. Hence $h_{t,n}(t)$ can be approximated by $n^{1/2}/\sqrt{2\pi}\sigma(t)$ and $h_{t,n}'/h_{t,n}(t)$ by $\sigma'/\sigma(t)$ each with errors of order $1/n$. The term of order $n^{-1/2}$ disappears since we are evaluating the density at the mean.

From this

$$\begin{aligned} f_n'/f_n(t) &= -nc'/c(t) - \sigma'/\sigma(t) + 0(1/n) \\ &= -n\alpha(t) - \sigma'/\sigma(t) + 0(1/n). \end{aligned} \tag{5.1}$$

To illustrate the behavior of $f_n'/f_n(t)$, we examine its behavior for the case of the uniform density on $[-1, 1]$ and for the extreme value density, $f(x) = \exp\{x - \exp(-x)\}$. For the uniform case, it is possible to compute the exact value of $f_n'/f_n$. The following Exhibit 5.1 compares the exact and approximate values. See Exhibit 3.7 and 3.8 for results on the density and distribution function.

| t | $n=5$ Exact | $n=5$ Approximate | $n=20$ Exact | $n=20$ Approximate |
|---|---|---|---|---|
| 0.00 | 0.0000 | 0.0000 | 0.0000 | 0.0000 |
| 0.05 | -0.6608 | -0.6546 | -3.9140 | -3.9150 |
| 0.10 | -1.3270 | -1.3230 | -5.8540 | -5.8560 |
| 0.20 | -2.6950 | -2.7270 | -11.9200 | -11.5200 |
| 0.30 | -4.1520 | -4.1680 | -18.4500 | -18.4500 |
| 0.40 | -5.7700 | -5.7890 | -25.8200 | -25.8200 |
| 0.50 | -7.6610 | -7.7420 | -34.6100 | -34.6200 |
| 0.70 | -13.5200 | -13.3300 | -63.0700 | -63.0500 |
| 0.90 | -40.0000 | -40.0000 | -190.0000 | -190.0000 |

**Exhibit 5.1**
Exact and approximate results for $f_n'/f_n(t)$
uniform observations.

From Exhibit 5.1, it is clear that even for $n = 5$, the approximation is very accurate over the whole range. The following plots (Exhibit 5.2, 5.3) demonstrate the approach to normality as $n$ increases. As $n$ increases, $f_n'/f_n$ becomes smoother, on the one hand, and only a smaller increasingly steep central part contains most of the mass of the distribution. To calculate the curves the graphs are plotted for values of $t$ corresponding to middle 99.8% of the density (i.e. tails of .001). In order to compute these percentiles, it is convenient to use the tail area approximation of Lugannani and Rice (1980) given by (3.27). It is worth noting that although the exact tail area requires an integration over the range $(t, \infty)$, we can obtain a very accurate approximation with the values only at the point $t$. This results in considerable saving of computational effort.

The graph for the uniform (Exhibit 5.2) only shows the upper part of the graph since $f_n$ is symmetric. It is clear that at $n = 40$, the graph shows very little deviation from a straight line indicating close agreement with the normal. We can argue that such a diagram makes it very easy to see how quickly a density is approaching its normal approximation. The second graph (Exhibit 5.3) for the extreme value density shows a density for the mean which is decidedly asymmetric at least for values of $n = 5$ and 10. However for $n = 40$, we have good agreement with the normal.

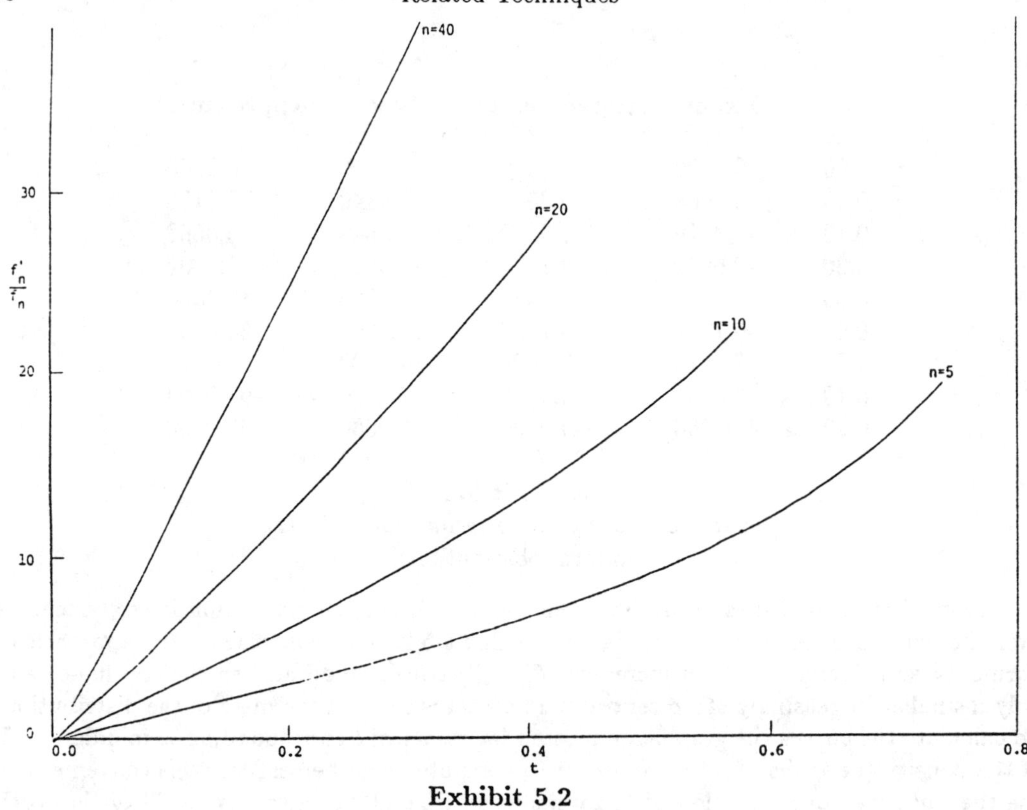


**Exhibit 5.2**
$f'_n/f_n$ for mean of uniform observations

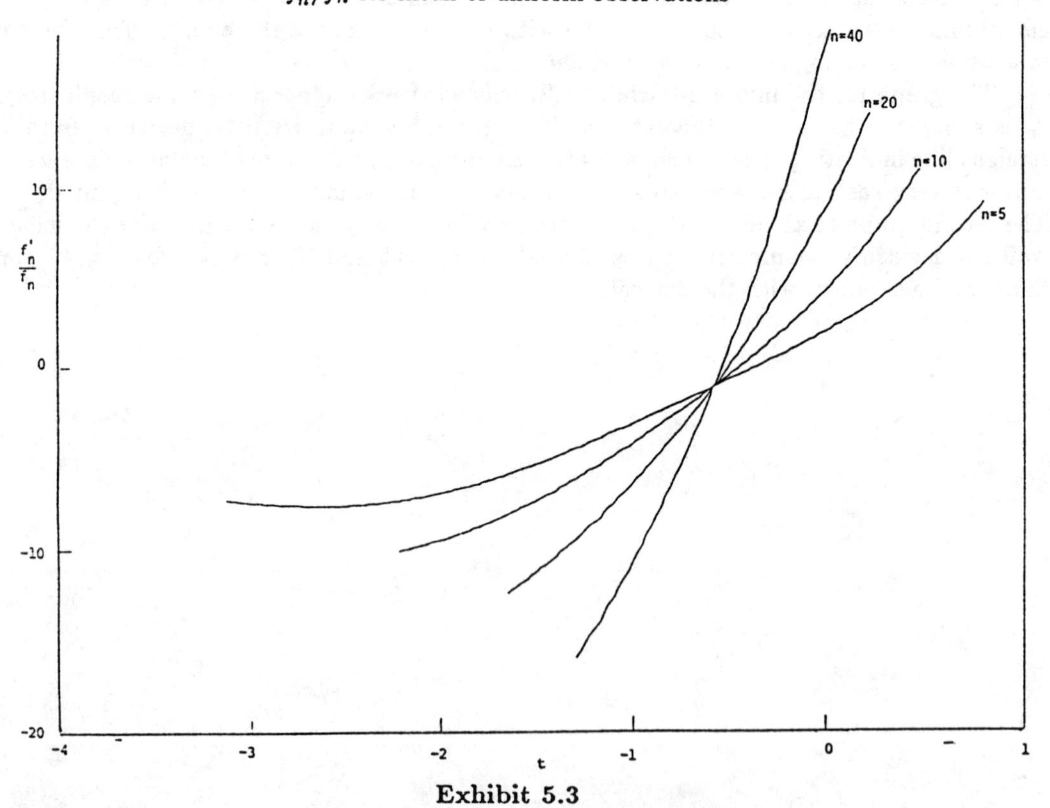


**Exhibit 5.3**
$f'_n/f_n$ for mean of extreme observations

In order to assess the quality of the approximation, we begin by considering the density of $n^{1/2}(\bar{X}-t)/\sigma(t)$ where we now assume that the $X_i$'s are distributed according to the conjugate density, $h_t(x)$. Denote this density of $n^{1/2}(\bar{X}-t)/\sigma(t)$ by $s_n(x;t)$. Now $s_n(0;t) = n^{-1/2}c_n(t)\sigma(t)f_n(t)$. But this implies that

$$\frac{f_n'}{f_n}(t) = -n\alpha(t) - \frac{\sigma'}{\sigma}(t) \quad \text{if} \quad \frac{\partial s_n(0;t)/\partial t}{s_n(0;t)} = 0.$$

Recalling that $s_n(0;t)$ is the density of a normalized sum at its expected value, we have an Edgeworth expansion as follows:

$$s_n(0;t) = (\pi/2)^{1/2}(1 + 0(1/n)).$$

The quality of the approximation will be determined by how closely the density of $n^{1/2}(\bar{X}-t)/\sigma(t)$ matches that of a standard normal in a neighborhood of $t$. Recent work by Field and Massam (1987) develops a diagnostic function based on this observation. The diagnostic function has some similarity to the diagnostic function for normality proposed by Efron (1981). As will be demonstrated in chapter 6, we can think of the small sample approximation being based on a local transformation to normality. Efron uses a similiar , but a less general transformation, as a means of constructing confidence intervals (cf. Efron (1987)). Section 6.3 gives a construction based on small sample approximations.

We now turn to a brief discussion of $\alpha(t)$. Recall that for the mean $\alpha(t)$ solves $\int(x-t)\exp\{\alpha(t)(x-t)\}f(x)dx = 0$. We note that $\alpha(t)$ uniquely determines $f$. For if $f_1$ and $f_2$ both give rise to the same $\alpha(t)$, then $\alpha(t) = c'(t)/c(t)$ which implies $\log c(t) = \int_\mu^t \alpha(s)ds$ so that $c(t)$ is the same for both $f_1$ and $f_2$. The relationship between $\alpha(t)$ and $c(t)$ can be obtained by differentiating the equation for $\alpha(t)$. From this

$$\int \exp(\alpha(t)(x-t))f_1(x)dx = \int \exp(\alpha(t)(x-t))f_2(x)dx.$$

Since these expressions are Laplace transforms, this implies the equality of $f_1$ and $f_2$.

To see the central limit theorem from this perspective, note that $\alpha(t) = t$ for the standard normal. Hence the central limit theorem requires

$$\lim_{n\to\infty} \alpha_n^*(t) = t$$

where $\alpha_n^*(t)$ corresponds to $\sqrt{n}(\bar{X}-\mu)/\sigma_0$, $E_f X_i = \mu$, $\sigma_0 = var_f X_i$. The first step is to express $\alpha_n^*(t)$ in terms of $\alpha(t)$ corresponding to $f$. $\alpha_n^*(t)$ must satisfy

$$E_f\left[\left(\frac{\sqrt{n}(\bar{X}-\mu)}{\sigma_0} - t\right)\exp\left(\alpha_n^*(t)\left(\frac{\sqrt{n}(\bar{X}-\mu)}{\sigma_0} - t\right)\right)\right] = 0$$

or

$$\frac{1}{\sigma_0\sqrt{n}}\sum_{i=1}^n E_f\left[\left(X_i - \mu - \frac{\sigma_0 t}{\sqrt{n}}\right)\exp\left(\frac{\alpha_n^*(t)}{\sigma_0\sqrt{n}}\left(X_i - \mu - \frac{\sigma_0 t}{\sqrt{n}}\right)\right)\right] \times$$

$$\prod_{j\neq i} E_f \exp\left(\frac{\alpha_n^*(t)}{\sigma_0\sqrt{n}}\left(X_j - \mu - \frac{\sigma_0 t}{\sqrt{n}}\right)\right) = 0.$$

But we can make the first term in the product equal to 0 by setting $\alpha_n^*(t)/\sigma_0\sqrt{n} = \alpha(\mu + \sigma_0 t/\sqrt{n})$. Therefore $\alpha_n^*(t) = \sigma_0\sqrt{n}\alpha(\mu + \sigma_0 t/\sqrt{n})$. Expanding $\alpha$ about $\mu$, we obtain

$$\alpha_n^*(t) = \sigma_0\sqrt{n}\alpha(\mu) + t\alpha'(\mu)\sigma_0^2 + \sigma_0^3 t^2\alpha''(\tilde{\mu}/2\sqrt{n}) \quad \text{for} \quad \mu < \tilde{\mu} \leq \mu + \sigma_0 t/\sqrt{n}.$$

We assume that $\alpha''(\tilde{\mu})$ is well behaved near $\mu$. Using the fact that $\alpha(\mu)=0$, $\alpha'(\mu)=1/\sigma_0^2$, we have $\lim_{n\to\infty}\alpha_n^*(t)=t$ as required for the central limit theorem.

Although we have chosen to focus on $\alpha(t)$ as the transform, it is also possible to view $\log c(t)(=\int_\mu^t \alpha(s)ds)$ as a transform. If we let $K(\alpha)\equiv\log E_f\exp(\alpha X)$ denote the cumulant generating function, then the Legendre transform $K^*(t)$ of $K(\alpha)$ is defined as

$$K^*(t)=\sup_\alpha\{\alpha t-K(\alpha)\}.$$

The maximizing value of $\alpha$ is obtained by solving

$$K'(\alpha)=t \qquad \text{or}$$

$$\int(x-t)\exp(\alpha)f(x)dx=0.$$

Hence

$$\begin{aligned}\alpha=\alpha(t) \quad \text{and} \quad K^*(t)&=\alpha(t)t-\log\int\exp(\alpha(t)x)f(x)\\&=\log\int\exp(\alpha(t)(x-t))f(x)dx\\&=-\log c(t),\\K^{*\prime\prime}(t)&=\alpha'(t)=1/K''(\alpha(t)),\end{aligned}$$

and the saddlepoint approximation can be written as

$$f_n(t)=(2\pi/n)^{1/2}(K^{*\prime\prime}(t))^{1/2}\exp(-nK^*(t))[1+O(1/n)].$$

A nice development of saddlepoint approximations using the Legendre transformation can be found in McCullagh (1987, Chapter 6).

Although $f_n'/f_n$ seems in many ways to be the most natural quantity to approximate, the arguments become awkward if we move from the mean, either to M-estimates or multi-dimensional problems (cf. Field and Hampel 1982). In these cases we end up approximating both $f_n$ and $f_n'$ and then taking the quotient of the results. Given the approximation to $f_n$, there seems to be no practical argument to do all the work to approximate $f_n'/f_n$. The fact that $f_n'/f_n$ gives us the correct integrating factor can be carried over to approximating $f_n$ by a renormalization of the approximation.

It is quite possible that $f_n'/f_n$ can be approximated directly in which case this would be an attractive alternative. This may involve a more geometric approach than the one we have been using. Insight in this direction could be very helpful in obtaining a deeper understanding of the mechanics of the approximation and could lead to new proofs and deeper understandings of the central limit theorems.

## 5.3. RELATIONSHIP TO EDGEWORTH EXPANSION AND LARGE DEVIATIONS

In this section, our formula for $f_n'/f_n$ and the saddlepoint method are compared to the classical methods of Edgeworth expansion (Cramer, 1946, pp. 229, 223, 133 etc.; Daniels,

1954) and large deviations (Richter, 1957; Feller, 1971; Cramer, 1938). In addition, from the formula for $f_n'/f_n$ a new variant of the classical methods is derived. Using this variant the connections between the methods is made very clear. The development follows closely that of Field and Hampel (1982, section 10).

To make comparisons easy, we consider the case of the arithmetic mean i.e. $\psi(x) = x$. Let $X_1, \cdots, X_n$ be independent observations from a density $f$ satisfying regularity conditions required for the classical expansions; for example, conditions 1 and 2 of Richter (1957) p. 208, which require that the moment generating function of $f$ exists in an interval and that $\sum_{i=1}^{n} X_i$ has a bounded density. Assume that $EX_i = 0$ and put $\text{var}X_i = \sigma^2$, $EX_i^3/\sigma^3 = \lambda_3$, $EX_i^4/\sigma^4 - 3 = \lambda_4$. Write $T_n = \bar{X}$ with density $f_n(t)$. Now $ET_n = 0$, $\text{var}T_n = \sigma^2/n$, $\lambda_3(T_n) = \lambda_3/\sqrt{n}$, $\lambda_4(T_n) = \lambda_4/n$.

Before proceeding, it is helpful to look at the situations to which these methods are directed. Both the Edgeworth and large deviation expansions approximate the density of $\sqrt{n}\bar{X}/\sigma$, which at a point $x$ equals $f_n(x\sigma/\sqrt{n})\sigma/\sqrt{n}$. In the Edgeworth expansion $x = 0(1)$ while in large deviations $x = 0(\sqrt{n})$. Writing $t = \sigma x/\sqrt{n}$, the methods can be compared as follows:

| | |
|---|---|
| $f_n'/f_n$ and saddlepoint : | $t = 0(1)$ |
| Large deviations up to order $k-2$ : | $t = 0(n^{-1/k}), \quad k > 2$ |
| Edgeworth : | $t = 0(n^{-1/2})$. |

In deriving $f_n'/f_n$ and the saddlepoint approximation at $t$, the underlying density is recentered around $t$ using a conjugate (or associated) distribution, $h_t(x) = c(t)\exp\{\alpha(t)(x - t)\}f(x)$ with $\int(x-t)h_t(x)dx = 0$. This centering is equivalent to a shift in the $f'/f$ space:

$$h_t'(x)/h_t(x) = f'(x)/f(x) + \alpha(t).$$

The Edgeworth expansion is used locally at 0 for each centered density, $h_t$, in both $f_n'/f_n$ and the saddlepoint approximation. In fact, the saddlepoint approximation, which is, except for a constant, the integrated version of $f_n'/f_n$, only uses the first term of the Edgeworth expansion, the normal approximation, at each $t$. It is remarkable that this simple device yields the very good accuracy even in the extreme tails that has been shown in the previous sections. On the other hand, the Edgeworth or large deviation expansion are not recentered. To take our comparisons a step further, only the local behavior of $f_n'/f_n$ and the saddlepoint approximation at $t = 0$ is considered.

Starting from formula (5.1), we have, for the arithmetic mean, $f_n'/f_n(t) = -n\alpha(t) - \beta(t) - \gamma(t)/n \cdots$ where $\alpha(t)$, $\beta(t) = \sigma'/\sigma(t)$, $\gamma(t)$ corresponds to terms of order $1/n$ which we will not need explicitly. Put $\alpha(t) = \alpha(0) + \sum_{v=1}^{\infty} \alpha^{(v)}(0)t^v/v!$, $\beta(t) = \beta(0) + \sum_{v=1}^{\infty} \beta^{(v)}(0)t^v/v!$ and $\gamma(t) = \gamma(0) + \sum_{v=1}^{\infty} \gamma^{(v)}(0)t^v/v!$. Recall that in both Edgeworth and large deviations $t \to 0$ as $n \to \infty$, so that these expansions make sense. By integrating, we obtain

$$\begin{aligned}\log f_n = \log f_n(0) &- n\alpha(0)t - \frac{n}{2}\alpha'(0)t^2 - \frac{n}{6}\alpha''(0)t^3 \cdots \\ &- \beta(0)t - \frac{1}{2}\beta'(0)t^2 - \frac{1}{6}\beta''(0)t^3 \\ &- \frac{1}{n}\gamma(0)t - \frac{1}{2n}\gamma'(0)t^2 \cdots.\end{aligned}$$

Write $\log f_n(0) = \log(\sqrt{n/2\pi\sigma^2})(1 + w_1/n + \cdots)$. Observing that $\alpha(0) = 0$, since $EX = 0$, we have

$$f_n(t) = \sqrt{n/2\pi\sigma^2}\exp\{w_1/n + \cdots\}\exp\{-n\alpha'(0)t^2/2\}\exp\{-\frac{n}{6}\alpha''(0)t^3 - \frac{n}{24}\alpha'''(0)t^4 \cdots - \beta(0)t - \frac{1}{2}\beta'(0)t^2 - \frac{1}{6}\beta''(0)t^3 - \frac{1}{n}\gamma(0)t - \frac{1}{2n}\gamma'(0)t^2 \cdots\}. \tag{5.2}$$

Note that the various terms can all be expressed in terms of $\sigma, \lambda_3, \lambda_4 \cdots$. This can be done directly by differentiating or from (5) and (6) in Richter (1957). However this re-expression of terms is easiest to see by means of comparison with the Edgeworth series which we do later in the section. Note that the infinite series (2.6) in Daniels (1954) has effectively $t = 0$ (after recentering) so that only the expansion of $f_n(0)$ remains in (5.2), i.e.$\sqrt{n/2\pi\sigma^2}e^{w_1/n + \cdots}$. In 2.6 the exponential is expanded to give $1 + w_1/n + \cdots$, so that for a finite piece of the series, negative approximated densities can result. However, the saddlepoint approximation, which ignores $w_1/n$ etc., is always positive and has been seen to be very accurate even in the extreme tails . With regards to the constant of integration, $\log f_n(0)$, numerical results indicate that accuracy can be improved over the expansion used above by evaluating this numerically as $\left[2\int_0^\infty f_n(t)dt\right]^{-1}$.

Key pages for the connection between large deviation and saddlepoint approximations are Richter (1957), p. 212 and 214. Note that in Richter's formulas, there are several misprints. On the bottom of p. 213, the formula for $I$, should have been $\varphi_3^2 t^6/2(3!)^2$ (the 2 is missing) and top of p.214, $\varphi_4(z_0)/8 - \frac{5}{24}\varphi_3^2(z_0)$ instead of $\varphi_4(z_0)/9 - 5\varphi_3^2(z_0)/12$.

A key formula for the connection between saddlepoint and Edgeworth approximations is (4.3) in Daniels (1954) where the Hermite polynomials differ from those in Cramer (1946) by a factor of $(-1)^n$.

Consider now the Edgeworth expansion which is an expansion for $n^{1/2}t = \text{constant} > 0$ (i.e. at each fixed multiple of the standard deviation of $T_n$). We proceed by expanding the exponents in (5.2). Remembering $nt^2 = \text{constant}$, groups of terms of equal order are:

| | |
|---|---|
| Constant, $nt^2$ | (together with $\sqrt{n}$, the normal approximation) |
| $nt^3, t$ | (skewness only in addition) |
| $nt^4, t^2, 1/n$ | (skewness and excess). |

The expansion of the exponentials in (5.2) up to this order yields

$$f_n(t) \cong \left(\frac{n}{2\pi\sigma^2}\right)^{1/2}\exp(-n\alpha'(0)t^2/2)\left\{1 - \frac{\alpha''(0)}{6}nt^3 - \beta(0)t - \frac{\beta'(0)}{2}t^2 + \frac{w_1}{n} - \frac{\alpha'''(0)}{24}nt^4 + \frac{\alpha''(0)\beta(0)}{6}nt^4 + \frac{\beta^2(0)}{2}t^2 + \frac{\alpha''(0)^2}{72}n^2t^6\right\}.$$

To match with the Edgeworth expansion, make the substitution $x = \sqrt{n}t/\sigma$. The Edgeworth expansion is an expression for fixed $x$ in powers of $n^{-1/2}$. From (2.9) we obtain the expansion for the standardized density

$$h_n(x) = \phi(x)\left\{1 + \frac{\lambda_3(T_n)}{3!}(x^3 - 3x) + \frac{\lambda_4(T_n)}{4!}(x^4 - 6x^2 + 3) + \frac{\lambda_3^2(T_n)}{72}(x^6 - 15x^4 + 45x^2 - 15)\right\}$$

where $\phi(x)$ is the standard normal density.

Since $\lambda_3(T_n) = \lambda_3/\sigma n$ etc., we obtain by matching terms $\alpha(0) = 0$, $\alpha'(0) = 1/\sigma^2$, $\alpha''(0) = -\lambda_3/\sigma^3$, $\alpha'''(0) = -\lambda_4/\sigma^4 + 3\lambda_3^2/\sigma^4$, $\beta(0) = \lambda_3/2\sigma$, $\beta'(0) = \lambda_4/2\sigma^2 - \lambda_3/\sigma^2$, $w_1 = \lambda_4/8 - 5\lambda_3^2/24$.

From Theorem 2 and (7) of Richter (1957), the first two terms of the large deviation expansion yield

$$f_n(t) \cong \sqrt{\frac{n}{2\pi\sigma^2}}e^{-nt^2/2\sigma^2}\exp\left\{\frac{\lambda_3}{6\sigma^3}nt^3 + \frac{\lambda_4}{4!\sigma^4}nt^4 - \frac{\lambda_3}{8\sigma^4}nt^4\right\}. \tag{5.3}$$

This expansion corresponds to an extreme case of asymptotic direction in which $n \to \infty$ first and then $t \to 0$ or the limiting case of $n^c t = \text{const}$ for $c \to 0$. This corresponds to keeping only the leading constant term and the expansion of $\alpha(t)$ and leads precisely to formula (5.3) above.

Hence the large deviation approximation for the density is nothing but the expansion of $\alpha(t)$ totalling ignoring the other terms in (5.2). Its value at $t = 0$ coincides with that of the saddlepoint approximation, but since for $t \neq 0$, it does not readjust $\sigma$ as the latter does (which amounts to keeping $\beta(t)$ in the local expression), even the full infinite large deviation series would (apart from a constant) correspond to using only the first order term in the integrated $f_n'/f_n$ approximation or the equivalent saddlepoint approximation. Results in Hampel (1973) show that this will give a poor numerical fit; the finite pieces of the series such as (5.3) above are only an approximation to this poor fit. The versions of large deviations for the cumulative, instead of the density, such as (6.23) in Feller (1971), probably contain the additional approximating error of the normal tail area by a function of the normal density and are likely to be still worse.

If we keep all terms of order $n^{1/2}t = \text{constant}$, then the full version of (5.2) up to this order is

$$f_n(t) \cong \sqrt{\frac{n}{2\pi\sigma^2}}\; e^{-nt^2/2\sigma^2}\exp\left\{\frac{\lambda_3}{6\sigma^3}nt^3 + \frac{\lambda_4}{24\sigma^4}nt^4 - \frac{\lambda_3}{8\sigma^4}nt^4 - \frac{\lambda_3}{2\sigma}t - \frac{\lambda_4}{4\sigma^2}t^2 + \frac{\lambda_3^2}{2\sigma^2}t^2 + \frac{\lambda_4}{8n} - \frac{5\lambda_3^2}{24n}\right\}. \tag{5.4}$$

The new formula (5.4) is what large deviations ought to be to give any hope of decent numerical results. It compares closely to Edgeworth, as the only difference is in the finite expansion of the exponent in Edgeworth. However the new formula can never be negative while the Edgeworth can. On the other hand, if $\lambda_4 - 3\lambda_3^2 \geq 0$, and not $\lambda_3 = \lambda_4 = 0$, formula (5.4) will eventually explode for large $|t|$, as will large deviations. Also the strict norming of the total probability, which Edgeworth expansions often pay for with negative densities, will be lost. But all this hardly matters if one cuts off for large $|t|$ when all these approximations, based on expansions at 0, will be bad anyway. Of greater interest is the behavior for small

$|t|$ and then it can be hoped that (5.4) is slightly better than Edgeworth. This has been supported by a numerical example with $X_i$ exponential, $n = 4$ when for $|x| = |\sqrt{n}t/\sigma| \leq 1.5$, the error was rougly halved and outside $|x| = 2$, both approximations were bad as is to be expected.

| t | Exact | "Edgeworth exponent (5.4) | % error | Edgeworth | % error | Normal | % error | Large deviations (5.3) | % error |
|---|---|---|---|---|---|---|---|---|---|
| -0.25 | 0 | .00168 | | -.03920 | | .0350 | | .0002 | |
| 0 | 0 | .04596 | | .02175 | | .1080 | | .0105 | |
| 0.25 | 0.24525 | .29544 | 20.5 | .02175 | 21.4 | .2590 | 5.6 | .0105 | -56.1 |
| 0.5 | .72179 | .70413 | -2.4 | .69231 | -4.1 | .4840 | -11.9 | .3848 | -46.7 |
| 0.75 | .89617 | .89126 | -.5 | .88857 | -.8 | .7042 | -21.4 | .6869 | -23.4 |
| 1 | .78147 | .78143 | -.005 | .78126 | -.02 | .7979 | 2.1 | .7979 | 2.1 |
| 1.25 | .56150 | .56358 | .4 | .56584 | .8 | .7042 | 25.4 | .7162 | 27.6 |
| 1.5 | .35694 | .36151 | 1.3 | .36968 | 3.6 | .4840 | 35.6 | .5371 | 50.5 |
| 1.75 | .20852 | .20305 | -2.6 | .20052 | -3.8 | .2590 | 24.2 | .3313 | 58.9 |
| 2 | .11451 | .08952 | -21.8 | .09373 | -18.1 | .1080 | -5.7 | .1507 | 31.6 |
| 2.5 | .03027 | .00340 | -88.7 | .04026 | 33.0 | .0088 | -70.9 | .0051 | -83.2 |
| 3 | .00708 | $1.10^{-6}$ | -100 | .00889 | 25.6 | .0003 | -95.8 | $1.10^{-6}$ | -100 |
| 3.5 | .00152 | $7.10^{-14}$ | -100 | .00045 | -70.4 | $3.10^{-6}$ | -100 | $4.10^{-14}$ | -100 |
| 4 | .00031 | $2.10^{-17}$ | -100 | $6.10^{-6}$ | -98.1 | 1.10-8 | -100 | $3.10^{-28}$ | -100 |
| 5 | .00001 | $4.10^{087}$ | -100 | $3.10^{-11}$ | -100 | $2.10^{-11}$ | -100 | $8.10^{-89}$ | -100 |

**Exhibit 5.4**

Density of the mean of 4 exponential for various approximations
$E\bar{X}_4 = 1,\ var\bar{X}_4 = 1/4,\ \lambda_3(\bar{X}_4) = 1,\ \lambda_4(\bar{X}_4) = 3/2.$

The approximation of this using $f_n'/f_n$ is exact everywhere while the saddlepoint approximation has a constant relative error of +2.1% everywhere.

## 5.4. APPROXIMATING THE DENSITY OF SUFFICIENT ESTIMATORS AND MAXIMUM LIKELIHOOD ESTIMATORS IN EXPONENTIAL FAMILIES

We consider the problem of approximating the density of sufficient estimators with the aim of relating the results to those of small sample asymptotics. The development is based closely on the important work by Durbin (1980a) and we will use his notation. Durbin assumes that we have a matrix of observations $\mathbf{y} = (y_1, \cdots, y_n)^T$ (not necessarily iid) where each $y_i$ is of dimension $\ell$ and has density

$$f(\mathbf{y}, \theta) = G(\mathbf{t}, \theta)H(\mathbf{y})$$

where $\mathbf{t}$ is m-dimensional and is the value of an estimate $\mathbf{T}_n$ of $\theta$.

Durbin assumes that a transformation $y_1, \cdots, y_n \rightarrow t_1, \cdots, t_m,\ u_{m+1}, \cdots, u_{m\ell}$ exists so that the density of $T_n, g(\mathbf{t}, \theta) = G(\mathbf{t}, \theta)H_1(\mathbf{t})$.

The basic equation for the first step comes from rewriting

$$f(\mathbf{y}, \theta) = g(\mathbf{t}, \theta)h(\mathbf{y}) \quad \text{where} \quad h(\mathbf{y}) = H(\mathbf{y})/H_1(\mathbf{t}).$$

Since this result holds for any $\theta$, we obtain

$$g(\mathbf{t},\theta_0) = \frac{f(\mathbf{y},\theta_0)}{f(\mathbf{y},\theta)} g(\mathbf{t},\theta). \tag{5.5}$$

The approximation results by making an appropriate choice of $\theta$ and then approximating the last term on the right hand side using an appropriate Edgeworth expansion.

Although Durbin considers 4 cases, we will focus on case 4, the most general, in order to facilitate comparisons. In this development, we assume $\mathbf{T}_n$ satisfies the conditions required for the expansion (28) of Durbin to hold. This general Edgeworth expansion requires that certain regularity conditions hold and that the cumulants are of the correct order. Conditions given in sections 4.2 or 4.5 are typical of what is required. The first step is to choose the value of $\theta$ on the right hand side of (5.5). This is done by using the value $\tilde{\theta}$ such that $E(\mathbf{T}_n|\theta) = \mathbf{t}$, i.e.

$$\int (\mathbf{u}-\mathbf{t}) G(\mathbf{u},\tilde{\theta}) H_1(\mathbf{u}) d\mathbf{u} = 0.$$

When we use an Edgeworth expansion for $g(\mathbf{u},\tilde{\theta})$, the odd order terms disappear since we are expanding at the mean of $\mathbf{T}_n$. Substituting the expansion gives (21) of Durbin (1980a) i.e.

$$g(\mathbf{t},\theta_0) = (n/2\pi)^{m/2} \left|D_n(\mathbf{t})\right|^{-1/2} f(\mathbf{y},\theta_0)/f(\mathbf{y},\tilde{\theta})$$

$$\times \left\{ 1 + \sum_{k=2}^{[r/2]} n^{-k+1} P_{n,2k}(0,t) + o(n^{-r/2+1}) \right\}$$

where $D_n(\theta) = nE\{\mathbf{T}_n - E(\mathbf{T}_n)\}\{\mathbf{T}_n - E(\mathbf{T}_n)\}^T$ and $P_{n,j}(\mathbf{x},\theta)$ is a generalized Edgeworth polynomial defined by

$$\frac{\left|D_n(\theta)\right|^{-1/2}}{(2\pi)^{m/2}} \exp\{-\mathbf{x}^T D_n^{-1}(\theta)\mathbf{x}\} P_{n,j}(\mathbf{x},\theta)$$

$$= \frac{1}{(2\pi)^m} \int_{R_m} \exp\left\{ -i\mathbf{z}^T\mathbf{x} - \frac{1}{2}\mathbf{z}^T D_n(\theta)\mathbf{z} \right\} \pi_{nj}(\mathbf{z},\theta) d\mathbf{z}$$

(see Durbin 1980a for details).
Using only the first term we have

$$g(\mathbf{t},\theta_0) = (n/2\pi)^{m/2} \left|D_n(\mathbf{t})\right|^{-1/2} f(\mathbf{y},\theta_0)/f(\mathbf{y},\tilde{\theta}) \{1 + 0(n^{-1})\}. \tag{5.6}$$

If we compare this approximation to (4.25) we can see similarities. The expression for $c^{-n}(\mathbf{t})$ has been replaced by $f(\mathbf{y},\theta_0)/f(\mathbf{y},\tilde{\theta})$ and $|detA||det\Sigma|^{-1/2}$ has been replaced by $|D_n(\mathbf{t})|$. To explore this further, recall that in small sample asymptotics, there are two steps; the recentering, followed by the use of the first term in an Edgeworth expansion, namely a normal approximation, Equation (5.5) corresponds to (4.23) relating the density under $f$ with the density under the conjugate distribution.

Using the notation of this section, the centering formula from chapter 4 would require (equations are written in univariate form for ease of notation)

$$\int (u-t) \exp\{\alpha(t)(u-t)\} G(u,\theta_0) H_1(u) du = 0.$$

Because of the sufficiency, Durbin uses

$$\int (u-t)G(u,\tilde{\theta})H_1(u)du = 0$$

to recenter where $\tilde{\theta}$ is analagous to $\alpha(t)$. As can be seen, the expression $\exp\{\alpha(t)(u-t)\}G(u,\theta)$ has been replaced by $G(u,\tilde{\theta})$ so that we are in effect using a different conjugate density. In each case, the recentering requires a rescaling of the new density. In chapter 4, this is done with $c^{-n}(t)$, while in (5.5) this is achieved with $f(y,\theta_0)/f(y,\tilde{\theta})$. In summary, the existence of a sufficient statistic enables us to recenter the density without the necessity of knowing the cumulant generating function. The approximation step is essentially the same in both the small sample asymptotics and Durbin's approach.

The major advance brought about by Durbin's approach is that approximation (5.6) does not require that the observations be independent. This makes the results very useful in time series settings. It is the existence of the sufficient statistic which provides the essential simplication to make the technique feasible for the case of dependent observations. In several of the cases Durbin considers, he assumes $E[\mathbf{T}_n|\theta] = \theta$ up to order $1/n$, so that $\tilde{\theta} = \mathbf{t}$ and the approximation becomes

$$\hat{g}(\mathbf{t},\theta_0) = (n/2\pi)^{m/2}\left|D(\mathbf{t})\right|^{-1/2} f(\mathbf{y},\theta_0)/f(\mathbf{y},\mathbf{t}).$$

We now turn to the special case of the exponential family with n independent observations $x_1,\cdots,x_n$. The density is given by

$$f(x,\theta) = \exp\{\theta u(x) - K(\theta) + d(x)\}.$$

The argument here is for a one-dimensional $\theta$ but it carries through routinely for the p-dimensional case. Although there are several approaches which all lead to the same result, we will use Theorem 4.1 to illustrate the connection of the small sample asymptotic approach to that of Durbin and of Barndorff-Nielsen.

We want to approximate the density of the maximum likelihood estimate $T_n$ (or $\hat{\Theta}$) at a point $t$ (or $\hat{\theta}$) with an underlying density $f(x;\theta_0)$. We will develop the formula in terms of $T_n$ and $t$ in order to remain consistent with the notation to date. The score function is $\psi(x,t) = u(x) - K'(t)$ and $E_\theta u(x) = K'(\theta)$. The conjugate density is $h_t(x) = c(t)\exp\{\alpha(t)(u(x)-K'(t)) + \theta_0 u(x) - K(\theta_0) + d(x)\}$. The centering condition requires that $E_{h_t}(u(x) - K'(t)) = 0$. If we choose $\alpha(t) = t - \theta_0$, then $h_t$ is exponential with parameter $t$ and $E_{h_t}u(x) = K'(t)$ as required . Now

$$c^{-1}(t) = \exp\{K(t) - tK'(t)\}/\exp\{K(\theta_0) - \theta_0 K'(t)\}$$

$$a(t) = E_{h_t}[\partial(u(x) - K'(t))/\partial t] = -K''(t)$$

$$\sigma^2(t) = E_{h_t}(u(x) - K'(t))^2 = K''(t).$$

From this, approximation (4.8) becomes

$$\hat{g}_n(t) = (n/2\pi)^{1/2}\frac{\exp\{nK(t) - ntK'(t)\}}{\exp\{nK(\theta_0) - n\theta_0 K'(t)\}}(K''(t))^{1/2}.$$

To compare with formula (13) of Reid (1988), we replace $t$ by $\hat{\theta}$ and note that in $L(\hat{\theta}) = \exp\{\hat{\theta}\Sigma u(x_i) - nK(\hat{\theta}) + \Sigma d(x_i)\}$, $\hat{\theta}$ satisfies the condition that $\Sigma u(x_i) = nK'(\hat{\theta})$ i.e. the maximum likelihood equation. Hence

$$\hat{g}_n(\hat{\theta}) = (n/2\pi)^{1/2}\{L(\theta_0)/L(\hat{\theta})\}(K''(\hat{\theta}))^{1/2} \tag{5.7}$$

with the convention that $\Sigma u(x_i) = nK'(\hat{\theta})$. If we replace $(K''(\hat{\theta}))^{1/2}$, the expected information by $j(\hat{\theta}) = -\partial^2 \log L(\theta)/\partial\theta^2\big|_{\theta=\hat{\theta}}$, then we obtain formula 13 of Reid (1988). (5.7) is often referred to as Barndorff- Nielsen's formula and appears in Barndorff-Nielsen (1980, 1983). It is the same as both the small sample approximation and the approximation given by Durbin for sufficient statistics. In order to put the results in historical perspective, Henry Daniels had noted this result expressed in (5.7) in a discussion of a paper by Cox (1958). It is interesting to see how many of the results used today come directly from the pioneering work of Henry Daniels (1954).

To conclude this section, we show the form of approximation (4.25) for a curved exponential family. The development is based on work by Hougaard (1985) and we use his setting and notation. Assume

$$\begin{aligned} f(x,\theta) &= \exp\{\theta'\mathrm{t}(x) - K(\theta) + h(x)\} \\ &= \exp\{\theta'\mathrm{t}(x) + h(x)\}/\phi(\theta). \end{aligned}$$

The parameter $\theta$ is a function $\theta(\beta)$ of a p-dimensional parameter $\beta$. We are interested in approximating the density of $\hat{\beta}$, the maximum likelihood estimate of $\beta$. This setting includes non-linear regression with normal errors, logistic regression and log-linear models. $\hat{\beta}$ is obtained as the solution of

$$n(\bar{t} - \tau(\theta(\beta))'d\theta/d\beta = 0 \quad \text{where} \quad \tau(\theta) = E_\theta \mathrm{t}(X) = K'(\theta).$$

The equation which $\alpha$ must satisfy is

$$\int \psi_r(x,\beta)\exp\{\Sigma\alpha_j\psi_j(x,\beta)\}f(x)dx = 0, \quad r = 1,\cdots,p$$

or equivalently

$$\int \frac{d}{d\alpha_r}\exp\{\Sigma\alpha_j\psi_j(x,\beta)\}f(x)dx = 0.$$

If we substitute, the left hand side becomes

$$\int \frac{d}{d\alpha_r}\exp\left[(\mathrm{t}(x) - \tau(\theta))'\frac{d\theta}{d\beta}\alpha + \mathrm{t}(x)'\theta_0 - K(\theta_0) + h(x)\right]dx.$$

Assuming we can interchange integration and differentiation, it is possible to carry out the integration to obtain

$$\frac{d}{d\alpha_r}\left[\exp\left(-\tau(\theta)'\frac{d\theta}{d\beta}\alpha\right)\exp\left(K\left(\theta_0 + \frac{d\theta}{d\beta}\alpha\right) - K(\theta_0)\right)\right].$$

To have this derivative zero, it suffices to have the derivative of the log equal to 0. i.e.

$$\left(\tau\left(\theta_0 + \frac{d\theta}{d\beta}\alpha\right)\right) - \tau(\theta(\beta))'d\theta/\partial\beta_r = 0.$$

Since the conjugate is exponential with parameter $\theta^* = \theta_0 + d\theta/d\beta\alpha$, it is straightforward to compute $c^{-1}$, $\Sigma$, $A$. The resulting approximation is

$$f_n(\beta) =(n/2\pi)^{p/2} \exp\left\{n\left(K(\theta_0) - K(\theta^*) - \tau(\theta)'\frac{d\theta}{d\beta}\alpha\right\}\right.$$
$$\left[\left|\frac{d\theta'}{d\beta}\frac{d^2K}{d\theta^2}\frac{d\theta}{d\beta} - \{\tau(\theta^*) - \tau(\theta)\}'\frac{d^2\theta}{d\beta^2}\right|\left|\frac{d\theta'}{d\beta}\frac{d^2K}{d\theta^{*2}}\frac{d\theta}{d\beta}\right|^{-1/2}\right]$$
$$\times\left[1 + 0(1/n)\right] \tag{5.8}$$

where $\theta^* = \theta_0 + d\theta/\partial\beta\alpha$ with $\alpha$ given by $(\tau(\theta^*) - \tau(\theta))'d\theta/d\beta = 0$.
In the last section of this chapter, this formula will be applied to an example. It is important to recall that (5.8) is the small sample approximation (4.25) applied to a curved exponential family. The form of the exponential leads to simplification in that the integration in (4.25) can be carried out explicitly. Result (5.8) is given in Hougaard (1985) as Theorem 1 which he in turn attributes to Skovgaard (1985).

## 5.5. CONDITIONAL SADDLEPOINT

Consider the situation where we require an approximation to the density of a statistic $T_1$ given that $T_2 = t_2$. In what follows we assume that we can approximate the density of $T = (T_1, T_2)$ as well as the density of $T_1$. The most direct way to proceed is to use a small sample approximation $\hat{g}_n(t_1, t_2)$ for the joint density, a small sample approximation $\hat{g}_n(t_2)$ for the density of $T_2$ and then divide the two approximations to give an approximation to the conditional density. In the literature, this approximation is referred to as the double saddlepoint approximation. To be specific we would choose $(\alpha_1(\mathbf{t}), \alpha_2(\mathbf{t}))$ in the joint conjugate to center $(T_1, T_2)$ at the point $\mathbf{t}$ and $\alpha(t_2)$ in the conjugate for $T_2$ to center $T_2$ at $t_2$. Note that it may be a non-trivial process to compute $\alpha(t_2)$ for the marginal density of $T_2$. If $T_1$ and $T_2$ are both means, the score functions can be solved independently of each other making $\alpha(t_2)$ easy to compute. In other cases, this is not usually the case and there is no clear way to proceed.

In the case where we have sufficient estimators, Durbin provides a method of approximating the conditional density (section 4, Durbin 1980a). Assume $\mathbf{T}$ is a sufficient estimate of $\theta$ with bias of order $n^{-1}$ at most and that $T = (T_1, T_2)$ and $\theta = (\theta_1, \theta_2)$. By sufficiency, the joint density of $(T_1, T_2)$ can be written as $g(t_1, t_2; \theta_1, \theta_2)$ and $f(x, \theta) = g(t_1, t_2; \theta_1, \theta_2)h(x)$.

Durbin considered the case where the conditional density of $T_1$ given $T_2$ depends only on $\theta_1$. Using arguments as in section 5.4, we have

$$g(t_1, t_2; \theta_{10}, \theta_2) = \frac{f(x; \theta_{10}, \theta_2)}{f(x; t_1, t_2)} g(t_1, t_2; t_1, t_2)$$

where $\theta_{10}$ is the particular value at which we require the approximation. Similarly, by sufficiency

$$g_2(t_2; \theta_{10}, \theta_2) = \frac{f(x; \theta_{10}, \theta_2)}{f(x; \theta_{10}, t_2)} g_2(t_2; \theta_{10}, t_2).$$

Dividing the equations yields

$$g(t_1|t_2, \theta_{10}) = \frac{f(x; \theta_{10}, t_2)}{f(x; t_1, t_2)} \frac{g(t_1, t_2; t_1, t_2)}{g_2(t_2; \theta_{10}, t_2)}.$$

Effectively the sufficiency and unbiasedness has enabled us to center the densities at $t_1$ and $t_2$. It remains only to replace $g(t_1, t_2; t_1, t_2)$ and $g_2(t_2; \theta_{10}, t_2)$ by their normal approximations to obtain an approximation with an error term of order $n^{-1}$; cf. also Skovgaard (1987).

Another case of interest is when we have an ancillary statistic. This case has been studied at length by Barndorff-Nielsen in several papers (cf. Barndorff-Nielsen 1983, 1984, 1986). To be specific, assume $T_1$ is an estimate of $\theta_1$ and we have a minimal sufficient statistic $(T_1, T_2)$ where $T_2$ is ancillary. Hence $f(x, \theta_1) = g(t, \theta_1)h(x) = g_1(t_1|t_2, \theta_1)g_2(t_2)h(x)$ since $T_2$ is ancillary. We can replace $\theta_1$ by $t_1$ in the expression above and then divide expressions to obtain as Durbin does,

$$g_1(t_1|t_2, \theta_1) = \frac{f(x, \theta_1)}{f(x, t_1)} g(t_1|t_2, t_1).$$

The final step is to approximate $g(t_1|t_2, t_1)$ by its normal approximation. One way to do this is to use the limiting variance $D(\theta_1)$ of $\sqrt{n}(T_1 - \theta_1)$ under the conditional distribution of $T_1$ given $T_2$. This gives us the approximation

$$g_1(t_1|t_2, \theta_1) = \left(\frac{n}{2\pi}\right)^{m_1/2} |D(t_1)|^{-1/2} \frac{f(x, \theta_1)}{f(x, t_1)} \{1 + 0(n^{-1})\} \tag{5.9}$$

where $m_1$ is the dimension of $\theta_1$. This is exactly expression (27) of Durbin (1980a).

To relate this formula to the extensive work of Barndorff-Nielsen, it is useful to rewrite (5.9) as

$$f_{\hat{\theta}|A}(\hat{\theta}|a; \theta) = c(\theta, a)|j(\hat{\theta})|^{1/2} \left\{L(\theta; \hat{\theta}, a)/L(\hat{\theta}; \hat{\theta}, a)\right\} \left\{1 + 0\left(\frac{1}{n}\right)\right\} \tag{5.10}$$

(cf. Reid 1988, formula 15).
Although the notation is different, the two formulas are the same except that in (5.10) the approximate density has been renormalized. The original work by Barndorff-Nielsen focussed on exponential families and transformation models and showed that formula (5.10) is exact in a number of cases.

McCullagh (1984) considers a fairly general situation and shows that formula (5.10) is generally valid (cf. (37) and the preceeding argument in McCullagh). However it is not easy to see how to construct the required second-order locally ancillary statistic A which is needed to carry out computations.

The simplest example where (5.10) is exact is that of the location/scale family. In this case the ancillary a is given by $\mathbf{a} = (a_1, \cdots, a_n)$, $a_i = (x_i - \hat{\mu})/\hat{\sigma}$ where $(\hat{\mu}, \hat{\sigma})$ is the maximum likelihood estimate of $(\mu, \sigma)$. Fisher (1934) showed that the density for $(\hat{\mu}, \hat{\sigma})$ given the ancillary a can be written as

$$f(\hat{\mu}, \hat{\sigma}|\mathbf{a}; \mu, \sigma) = c_0(\mathbf{a})\hat{\sigma}^{n-2} f(\mathbf{x}; \mu, \sigma) \tag{5.11}$$

where the $\mathbf{x}$ on the right hand side is expressed as

$$\mathbf{x} = (\hat{\mu} + \hat{\sigma}a_1, \cdots, \hat{\mu} + \hat{\sigma}a_n).$$

To see how this is related to formula (5.10), note that $L(\mu, \sigma; \mathbf{x}) = f(\mathbf{x}; \mu, \sigma)$ so that $L(\hat{\theta}; \hat{\theta}, \mathbf{a}) = \hat{\sigma}^{-n} f(\mathbf{a}; 0, 1)$. Also $|j(\hat{\theta})| = D(\mathbf{a})\hat{\sigma}^{-4}$ where

$$D(\mathbf{a}) = \{\Sigma g''(a_i)\}\{n + \Sigma a_i^2 g''(a_i)\} - \{\Sigma a_i g''(a_i)\}^2$$

with $g(\mathbf{x}) = -\log f(\mathbf{x})$. Hence

$$|j(\hat{\theta})|^{1/2}\{L(\theta;\hat{\theta},\mathbf{a})/L(\hat{\theta};\hat{\theta},\mathbf{a})\} = D^{1/2}(\mathbf{a})\hat{\sigma}^{n-2}f(\mathbf{x};\mu,\sigma)/f(\mathbf{a};0,1)$$

where $\mathbf{x}$ is expressed as above.
Formula (5.11) can be written as

$$c(\mathbf{a})\hat{\sigma}^{n-2}f(\mathbf{x};\mu,\sigma) \quad \text{where} \quad c(\mathbf{a}) = c_0(\mathbf{a})f(\mathbf{a};0,1)/D^{1/2}(\mathbf{a})$$

and we see that for location/scale (5.10) is exact. The reader is referred to Barndorff-Nielsen for other examples where exactness holds.

It is interesting to consider the relationship between the approximation conditional on an ancillary statistic and the unconditional approximation. Although this relationship is not clear the situation of observations from a normal with mean $\theta$ and variance $b^2\theta^2$ with $b^2$ known is one in which computations could be carried out relatively easily. The conditional formula is given in Reid (1988, cf formula 17) and is known to be exact. The unconditional formula can be worked out with some effort. $\alpha(t)$ can be computed explicitly so it appears the approximation can be given explicitly and the two approximations compared.

There is a need for more research to establish connections between the basic small sample approximations and the work of Barndorff-Nielsen.

## 5.6. NONTRIVIAL APPLICATION IN THE EXPONENTIAL FAMILY: LOGISTIC REGRESSION

In this section, we consider the example of logistic regression through the origin. Since the model falls within the exponential family, we might expect the approximation to be very straightforward. However, as we shall see, there are some complications in obtaining useful results. Consider the usual set-up for logistic regression through the origin with

$$P[Y=1] = e^{\beta x}/(1+e^{\beta x}).$$

We want to approximate the density of $\hat{\beta}$, the maximum likelihood estimate of $\beta$. In this example, we consider $X$ to be random with a density $f(x)$. The situation in which $x$ is considered fixed has no essential differences except for some increased complexity in notation.

Assume that $X$ has a density $f(x)$. To follow the notation developed to date, $\psi(y,x,\beta)$ is the derivative of the log likelihood function and is given by

$$\psi(y,x,\beta) = yx - x\exp(x\beta)/(1+\exp(x\beta)).$$

A direct approach is to evaluate $f_n(t)$ using the approximation (4.8) for M-estimates. It is straightforward to verify that $\alpha(t) = t-\beta$ and the approximation can easily be computed.

If these computations are done and the results compared to the asymptotic results, there are some clear discrepancies. For instance if $f(x)$ is normal and $\beta = 0$, the variance for a fixed $n$ can be computed from $f_n(t)$ and can be approximated on the basis of the asymptotic variance. The results are as follows:

| | n | | | | |
|---|---|---|---|---|---|
| **Variance of $\hat{\beta}$** | 5 | 10 | 20 | 40 | 100 |
| based on $f_n(t)$ | 19.6 | 2.2 | .32 | .12 | .04 |
| based on asymptotic variance | .8 | .4 | .2 | .1 | .04 |

It's clear that in the range of interest, namely 5–20, the approximation is giving results in which the density is much longer-tailed than is suggested by the asymptotic theory. To understand this problem, it is necessary to look at the equation for solving $\beta$ i.e.

$$\Sigma y_i x_i - \Sigma x_i \exp(\beta x_i)/(1+\exp(\beta x_i)) = 0.$$

If $y_i = 1$ for all positive $x_i$'s and 0 for all negative $x_i$'s , then $\beta = \infty$ is the maximum likelihood estimate. The observed long tail of the approximation for small to moderate $n$ is due to a positive mass at $\pm\infty$. $f_n(t)$ is a smooth approximation to a mixture of a continuous density and point masses at $\pm\infty$.

In practice, the density of interest is the density of $\hat{\beta}$ conditional on $\hat{\beta}$ finite since the experimenter will only be interested in making inferences about $\beta$ in the case that $\hat{\beta}$ is finite. We now proceed with adapting our basic approximation to handle this case.

To obtain an approximation, we need to find an appropriate conjugate density and verify a centering lemma. As a first step, consider the moment generating function $\hat{\beta}$ and $\Sigma\psi(y, x, \beta)$ conditional on $\hat{\beta}$ finite. The true value of $\beta$ will be denoted as $\beta_0$. For ease of discussion, we assume that all the $x_i$'s are greater than or equal to 0. It then follows that

$$\hat{\beta} \text{ finite } \Leftrightarrow 1 \le \sum_{i=1}^{n} y_i \le n-1.$$

Also

$$\begin{aligned} P(\hat{\beta} \text{ finite}) &= 1 - \left(\int \exp(\beta_0 x)/(1+\exp(\beta_0 x))f(x)dx\right)^n \\ &\quad - \left(\int 1/(1+\exp(\beta_0 x))^n f(x)dx\right)^n \\ &= h(\beta_0) \quad \text{(say)}. \end{aligned}$$

The conditional density of $(X_i, Y_i)$, $i = 1, \cdots, n$ given $\hat{\beta}$ finite is

$$\exp\left(\beta \sum_{i=1}^{n} x_i y_i\right) \Big/ \left[\prod_{i=1}^{n}(1+\exp(\beta x_i))\right] P(\hat{\beta} \text{ finite})$$

$$\text{if } 1 \le \sum_{i=1}^{n} y_i \le n-1$$

and 0 otherwise.

The conjugate density at a point $\beta$ in this case will be a joint density in $n$ dimensions given by:

$$h_\beta(\mathbf{x}, \mathbf{y}) = c^n(\beta)\exp\left\{\beta_0 \sum_{i=1}^{n} y_i x_i + \alpha \sum_{i=1}^{n} \psi(y_i, x_i, \beta)\right\} \prod_{i=1}^{n} f(x_i)$$

$$I_{[1,n-1]}\left(\sum_{i=1}^{n} y_i\right) \Big/ \prod_{i=1}^{n}(1+\exp(\beta_0 x_i))$$

where $I_A$ is the indicator function for the set $A$. $\alpha$ must satisfy the condition

$$E_{h_\beta}\left[\sum_{i=1}^{n}\psi(Y_i,X_i,\beta)\right]=0. \tag{5.12}$$

Note that to solve (5.12) in its current form requires an n-dimensional summation (over $y_i$'s) and an n-dimensional integration (over $x_i$'s) and so is not computationally feasible. We will show how (5.12) can be simplified to obtain a computationally manageable form.

By using a proof very similar to that given in section 4.5 for the centering result (4.23), it can be shown that a similar result holds in this case, namely

$$f_n(\beta)=c^{-n}(\beta)h_{\beta,n}(\beta)$$

where $f_n$ represents the density of $\hat{\beta}$ under joint density of $X$ and $Y$ given by $f(x)$ and $\beta_0$ and $h_{\beta,n}(\cdot)$ represents the density of $\hat{\beta}$ under $h_\beta$. The densities are all conditional on $\hat{\beta}$ finite.

We now simplify (5.12). To begin let

$$g_\beta(x,y)=c(\beta)\exp\{\beta_0 yx+\alpha\psi(y,x,\beta)\}f(x)/(1+\exp(\beta_0 x_i)).$$

Now

$$h_\beta(\mathbf{x},\mathbf{y})=\prod_{i=1}^{n}g_\beta(x_i,y_i)I_{[1,n-1]}\left(\sum_{i=1}^{n}y_i\right).$$

To simplify notation, let $\Sigma'$ denote the sum over all vectors $\mathbf{y}$ of 0's and 1's with $1\le\sum_{i=1}^{n}y_i\le n-1$. Now (5.12) becomes

$$\Sigma'\int_{\cdots}\int\sum_{i=1}^{n}\psi(x_i,y_i,\beta)g_\beta(x_i,y_i)\prod_{j\neq i}g_\beta(x_j,y_j)d\mathbf{x}=0.$$

Let $\ell(y,\beta)=\int g_\beta(x,y)dx=\int\exp\{\beta_0 y+\alpha\psi(y,x,\beta)\}f(x)/(1+\exp\beta_0 x)dx$. We can now write our equation as

$$\Sigma'\sum_{i=1}^{n}\int\psi(x_i,y_i,\beta)g_\beta(x_i,y_i)\prod_{j\neq i}\ell(y_i,\beta)dx_i=0$$

or rearranging the summation, obtain

$$\sum_{r=1}^{n-1}\binom{n}{r}\Bigg[\int\sum_{i=1}^{r}\psi(x_i,1,\beta)g_\beta(x_i,1)(\ell(1,\beta))^{r-1}(\ell(0,\beta))^{n-r}dx_i$$
$$+\int\sum_{i=r+1}^{n}\psi(x_i,0,\beta)g_\beta(x_i,0)(\ell(1,\beta))^{r}(\ell(0,\beta))^{n-r-1}dx_i\Bigg]=0.$$

Simplifying, (5.12) can be written as:

$$\sum_{r=1}^{n-1}(\ell(1,\beta))^{r-1}(\ell(0,\beta))^{n-r-1}\binom{n}{r}\Bigg[r\ell(0,\beta)\int\psi(x,1,\beta)g_\beta(x,1)dx$$
$$+(n-r)\ell(1,\beta)\int\psi(x,0,\beta)g_\beta(x,0)dx\Bigg]=0.$$

Now instead of having to evaluate an n-dimensional integral, we have to evaluate four one-dimensional integrals making the computation of $\alpha(\beta)$ straightforward. The approximation to the density of $\hat{\beta}$ conditional on $\hat{\beta}$ finite, $f_n(\beta)$, is then given by

$$g_n(\beta) = \left(\frac{n}{2\pi}\right)^{1/2} c^{-n}(\beta)a(\beta)/\sigma(\beta)$$

where $\sigma^2(\beta)$ has the same form as the left-hand side of (5.12) with $\psi^2$ instead of $\psi$ and a division by $P(\hat{\beta}$ finite) and $a(\beta)$ replaces $\psi$ by $\partial\psi/\partial\beta$ and similarly includes a division by $P(\hat{\beta}$ finite). Finally

$$c^{-1}(\beta) = \sum_{r=1}^{n-1} \binom{n}{r} (\ell(\beta,1))^r (\ell(\beta,0))^{n-r} / P(\hat{\beta} \text{ finite}).$$

The extension of this approximation to higher dimension can be made if we retain the condition that the $x_i$'s are positive. If the $x_i$'s can be both positive and negative, the condition that $\hat{\beta}$ be finite involves both $x_i$ and $y_i$ and it becomes more difficult to simplify (5.12) to involve only one-dimensional integrals.

# 6. APPROXIMATE CONFIDENCE INTERVALS IN MULTIPARAMETER PROBLEMS

## 6.1. INTRODUCTION

As has been noted in the previous chapters, a limiting aspect of the saddlepoint approximation has been the computational complexity in higher dimensions. In a problem with $p$ parameters, we are often interested in approximating the marginal density of the estimate of one of the parameters or the density of some test statistics. We could in principle use the approximation to obtain the density on a grid in p-dimensions and then integrate over appropriate regions of p-dimensional space. However, at each grid point, we have to solve a non-linear system of $p$ equations to obtain $\alpha$. If $p$ is three or more this is not really a feasible method numerically.

In this chapter, we introduce a technique developed by Tingley and Field (1990) designed to overcome this problem. In fact the approach to be presented will be based on a nonparametric bootstrap and will allow us to obtain correct one or two-sided second order confidence intervals without specifying an underlying density. One of the tools that is essential is the tail area approximation due to Lugannani and Rice (1980). This approximation eliminates the integration of the approximate density in calculating tail areas.

The next section discusses the tail area approximation while the third section demonstrates how to compute confidence intervals which are both robust and nonparametric.

## 6.2. TAIL AREA APPROXIMATION

Up to this point, we have developed approximations for densities of estimates. However in many situations, there is more interest in approximating the cumulative distribution. In particular, for confidence intervals and testing procedures, it is the tail area of the distribution which is of interest. In this section we will develop a tail area approximation for the case of the univariate mean. This approximation will then be used in the next section to construct confidence intervals in the multiparameter situation. The tail area approximation is based on uniform asymptotic expansions and was developed for tail areas by Lugannani and Rice (1980). Both Daniels (1987) and Tingley (1987) have placed the result in the context of small sample or saddlepoint approximations. Our development is similar to that of Daniels and Tingley but with some notational changes.

Consider the situation where we have $n$ independent, identically distributed random variables, $X_1, X_2, \cdots, X_n$ and we want to approximate $P(\bar{X} \geq x_0)$ for some point $x_0$. Based on our previous approximation for the density of $\bar{X}$ we could approximate the upper tail area by $\int_{x_0}^{\infty} kc^{-n}(x)/\sigma(x)dx$ where $k$ is the normalizing constant. In order to evaluate this integral, we have to evaluate $c(x)$ and $\sigma(x)$ (and hence $\alpha(x)$) over a grid of points from $x_0$ to $\infty$. As noted by Daniels (1987), the integration can be made simpler by a change of variables. The upper tail area can be written as $\int_{x_0}^{\infty} kc^{-n}(\alpha(x))/\sigma(\alpha(x))dx$. The monotone transformation $y = \alpha(x)$, gives the integral $\int_{\alpha(x_0)}^{\infty} kc^{-n}(y)\sigma(y)dy$ which avoids the necessity of computing the saddlepoint $\alpha(x)$ for every ordinate. Related work by Robinson (1982) gives a tail area approximation based on the Laplace approximation to the integral above.

The Lugannani-Rice tail area approximation will be based on the values of $c(x_0)$ and $\sigma(x_0)$ so we only require one evaluation of $\alpha$, namely $\alpha(x_0)$. This is a considerable simplication in computation and, as we shall see, gives remarkable accuracy.

We assume $X_i$ has a density $f$, a mean of 0 and a cumulant generating function $K(t)$ defined by

$$e^{K(t)} = E(e^{tX_1}).$$

The Fourier inversion formula gives the density of $\bar{X}$ as

$$f_n(x) = \frac{n}{2\pi} \int_{-\infty}^{\infty} e^{n(K(iu)-iux)} du.$$

Reversing the order of integration gives

$$P(\bar{X} \geq x_0) = \int_{x_0}^{\infty} \frac{n}{2\pi} \int_{-\infty}^{\infty} e^{n(K(iu)-iux)} du dx$$

$$= \frac{n}{2\pi} \int_{-\infty}^{\infty} e^{n(K(iu)-iux_0)} du/iu$$

$$= \frac{n}{2\pi i} \int_{-i\infty}^{i\infty} e^{n(K(z)-zx_0)} dz/z. \tag{6.1}$$

We denote the solution of the equation, $K'(z) = x_0$ by $\alpha$. Note that $\alpha$ is the saddlepoint and in the notation of the previous chapters would be denoted by $\alpha(x_0)$. The next step is to make a change of variables from $z$ to $t$ where

$$K(z) - zx_0 = t^2/2 - \gamma t + \rho.$$

$\gamma$ and $\rho$ are chosen so that $t = \gamma$ is the image of the saddlepoint $z = \alpha$ and the origin is preserved. This implies that $\rho = K(0) = \log \int f(x)dx = 0$ and $-\gamma^2 = 2(K(\alpha) - \alpha x_o)$. $\gamma$ takes the same sign as $\alpha$ i.e. $\gamma = sgn(x_0)\sqrt{-2(K(\alpha) - \alpha x_0)}$.

As we will show later on in the section, the choice of this transformation implies that a local normal approximation is being used. Under this transformation, (6.1) becomes

$$P(\bar{X} \geq x_0) = \frac{n}{2\pi i} \int_{b-i\infty}^{b+i\infty} e^{n(t^2/2-\gamma t)} G_0(t) dt/t$$

where $G_0(t) = (t/z)(dz/dt)$.

The effect of the transformation is to take the term which must be approximated out of the exponent where errors can grow very rapidly into the main part of the integrand. In order to obtain the terms in the expansion that we need, $G_0(t)$ is replaced by a linear approximation as follows. To begin, write

$$G_0(t) = a_0 + a_1 t + t(t-\gamma)H_0(t).$$

From this representation, $a_0 = G_0(0)$ and $a_1 = (G_0(0) - G_0(\gamma))/\gamma$. We now have to evaluate $G_0(0)$ and $G_0(\gamma)$. $G_0$ has removable singularities at both these points.

Starting with $K(z) - zx_0 = t^2/2 - \gamma t$, we obtain

$$\frac{dz}{dt} = \frac{t-\gamma}{K'(z) - x_0} \quad \text{so that} \quad \lim_{t\to 0} \frac{dz}{dt} \text{ exists.}$$

Hence

$$\lim_{t\to 0} G_0(t) = \lim_{t\to 0} \frac{t}{z} \lim_{t\to 0} \frac{dz}{dt} = \lim_{t\to 0} \frac{1}{dz/dt} \lim_{t\to 0} \frac{dz}{dt} = 1$$

so that $a_0 = 1$. Also

$$\lim_{t\to\gamma} G_0(t) = \lim_{t\to\gamma} \frac{t}{z} \lim_{t\to\gamma} \frac{dz}{dt} = \frac{\gamma}{\alpha(K''(\alpha))^{1/2}}$$

since

$$\lim_{t\to\gamma} \left(\frac{dz}{dt}\right) = \lim_{t\to\gamma} \frac{(t-\gamma)}{K'(z) - x_0} = \lim_{t\to\gamma} \frac{-1}{K''(z)\frac{dz}{dt}}$$

and

$$\lim_{t\to\gamma} \frac{dz}{dt} = \frac{1}{(K''(\alpha))^{1/2}}$$

so that

$$a_1 = -\frac{1}{\gamma} + \frac{1}{\alpha(K''(\alpha))^{1/2}}.$$

Approximating $G_0(t)$ by $a_0 + a_1 t$, we obtain

$$P(\bar{X} \geq x_0) \approx \frac{n}{2\pi i} \int_{b-i\infty}^{b+i\infty} \frac{1}{t}\, e^{-t^2/2-\gamma t} dt$$

$$+ \frac{n}{2\pi i}\left(-\frac{1}{\gamma} + \frac{1}{\alpha(K''(\alpha))^{1/2}}\right) \int_{b-i\infty}^{b+i\infty} e^{t^2/2-\gamma t} dt$$

$$\approx 1 - \Phi(\sqrt{n}\gamma) + \frac{e^{\sqrt{n}\gamma}}{\sqrt{2n\pi}}\left\{\frac{1}{\alpha(K''(\alpha))^{1/2}} - \frac{1}{\gamma}\right\}.$$

The next step is to rewrite the formula in more familiar terms.

$$\gamma^2 = -2(K(\alpha) - \alpha x_0) = -2\left(\log \int e^{\alpha(x-x_0)} f(x) dx\right) = 2 \log c(x_0)$$

and $\gamma = sgn(x_0)\sqrt{2\log c(x_0)}$. Similarly $K''(\alpha) = \sigma^2(\alpha)$. The approximation for the tail area becomes

$$P(\bar{X} \geq x_0) = 1 - \Phi(sgn(x_0)\sqrt{2n \log c(x_0)})$$

$$+ \frac{c^{-n}(t)}{\sqrt{2\pi}}\left\{\frac{1}{\alpha(x_0)\sigma(x_0)\sqrt{n}} - \frac{1}{sgn(x_0)\sqrt{-2n \log c(x_0)}}\right\}$$

$$\text{for } x_0 > E(X). \tag{6.3}$$

Daniels (1987) shows that the formula is exact if we add an error term $0(n^{-3/2})$ in the curly brackets of the second term in (6.3) (cf. Daniels formula (4.9)).

To demonstrate the accuracy of the approximation, we can consider situations in which the small sample approximation for the mean is exact. Then the error in the tail area when computed with (6.3) can be attributed to the error of the tail area approximation. The following numerical results in Exhibit 6.1 are taken from Daniels (1987) and Field and Wu (1988).

| | $n = 1$ | | | $n = 5$ | |
|---|---|---|---|---|---|
| $x_0$ | **Exact** | (6.3) | $x_0$ | **Exact** | (6.3) |
| .5 | .6065 | .6043 | .2 | .99634 | .99633 |
| 1 | .3679 | .3670 | .6 | .8153 | .8152 |
| 3 | .0498 | .0500 | 1 | .4405 | .4405 |
| 5 | $.0^2674$ | $.0^2681$ | 2 | .0293 | .0293 |
| 7 | $.0^3912$ | $.0^2926$ | 3 | $.0^3857$ | $.0^3858$ |
| 9 | $.0^3123$ | $.0^3126$ | 4 | $.0^4169$ | $.0^4170$ |
| | | | 5 | $.0^6267$ | $.0^6268$ |

**Exhibit 6.1a**

Tail area for mean under exponential
$f(x) = e^{-x}$, $x \geq 0$, $K(z) = -\log(1-z)$

| | $n = 3$ | | | $n = 5$ | |
|---|---|---|---|---|---|
| $x_0$ | **Exact** | (6.3) | $x_0$ | **Exact** | (6.3) |
| .33 | .9645 | .9638 | .2 | $.9^4466$ | $.9^4460$ |
| .67 | .6782 | .6724 | .6 | .8334 | .8315 |
| 1 | .3927 | .3848 | 1 | .4147 | .4108 |
| 1.67 | .1156 | .1108 | 2 | .0378 | .0328 |
| 3.33 | $.0^2548$ | $.0^2505$ | 4 | $.0^3148$ | $.0^3141$ |
| 6.67 | $.0^4174$ | $.0^4155$ | 5 | $.0^5994$ | $.0^5937$ |

**Exhibit 6.1b**

Tail area for mean under inverse normal
$f(x;\mu) = \mu \exp\{-(x-\mu)^2/2x\}/(2\pi)^{1/2}x^{3/2}$
$K(z) = \mu(1-(1-z)^{1/2})$, $\mu = 1$

As can be seen the relative error introduced by (6.3) is small with the maximum being just above 10% in the tables above. This demonstrates that the tail area approximation works remarkably well for small sample sizes and in the extreme tails. The following example for chi square random variables uses (6.3) iteratively to compute the upper percentiles for $x_5^2/df$. We used the tail area routine from $S$ to check our results in this case. $S$ uses an algorithm developed by Goldstein (1973) to approximate the tail area. In that paper there are estimates of the maximum relative error up to an upper tail area of .0005. For the situation with $n = 5$, the following results in Exhibit 6.2 were obtained.

| Upper tail area | .05 | $10^{-2}$ | $10^{-3}$ | $10^{-4}$ | $10^{-5}$ | $10^{-6}$ | $10^{-7}$ |
|---|---|---|---|---|---|---|---|
| Computed from $S$ | 2.2141 | 3.0173 | 4.1030 | 5.1490 | 6.1697 | 7.1616 | 8.1056 |
| Computed from (6.3) | 2.2150 | 3.0186 | 4.1049 | 5.1514 | 6.1740 | 7.1807 | 8.1760 |

**Exhibit 6.2**

Percentiles for $\chi_5^2/df$

The maximum relative errors for the algorithm used by $S$ are $10^{-2}$ for 5 degrees of freedom for tail areas up to $5 \times 10^{-4}$. For the tail areas in this range, we observe relative errors less than $4 \times 10^{-4}$, implying that approximation (6.3) again works very well.

As a final example we consider an example where the small sample approximation is not exact so that errors arise both from this and the tail area approximation; see Exhibit 6.3.

| | $n=1$ | | | $n=5$ | |
|---|---|---|---|---|---|
| $x_0$ | **Exact** | (6.3) | $x_0$ | **Exact** | (6.3) |
| .2 | .4 | .3838 | .2 | .2250 | .2249 |
| .4 | .3 | .2750 | .4 | .0620 | .0618 |
| .6 | .2 | .1791 | .6 | $.0^2833$ | $.0^2824$ |
| .6 | .1 | .0948 | .8 | $.0^3260$ | $.0^3255$ |

**Exhibit 6.3**
Tail area for mean under uniform
$f(x) = 1/2, -1 \le x \le 1, K(z) = \log(\sinh z/z)$

For this example, the tail area approximation works very well and the results from (6.3) are very close to those obtained by numerical integration.

If we consider the situation for robust location with a monotone score function, Daniels (1983) demonstrates that formula (6.3) remains valid. The following Exhibit 6.4 again demonstrates that (6.3) gives the same accuracy as numerical integration.

| | **5% Contaminated normal** | | | | **Cauchy** | | | |
|---|---|---|---|---|---|---|---|---|
| | **t** | **Exact** | **Integrated saddlepoint** | **Lugannani-Rice** | **t** | **Exact** | **Integrated saddlepoint** | **Lugannani-Rice** |
| | 0.1 | .46331 | .46229 | .46282 | 1 | .25000 | .28082 | .28197 |
| | 1.0 | .17601 | .18428 | .18557 | 3 | .10242 | .12397 | .13033 |
| n=1 | 2.0 | .04674 | .07345 | .07082 | 5 | .06283 | .08392 | .09086 |
| | 2.5 | .03095 | .06000 | .05682 | 7 | .04517 | .06484 | .07210 |
| | 3.0 | .02630 | .05520 | .05190 | 9 | .03522 | .05327 | .06077 |
| | 0.1 | .42026 | .42009 | .42024 | 1 | .11285 | .11458 | .11400 |
| | 1.0 | .02799 | .02779 | .02799 | 3 | .00825 | .00883 | .00881 |
| n=5 | 2.0 | .00414 | .00413 | .00416 | 5 | .00210 | .00244 | .00244 |
| | 2.5 | .00030 | .00043 | .00043 | 7 | .00082 | .00105 | .00104 |
| | 3.0 | .00018 | .00031 | .00031 | 9 | .00040 | .00055 | .00055 |
| | 0.1 | .39403 | .39393 | .39399 | 1 | .05422 | .05447 | .05427 |
| | 1.0 | .00538 | .00535 | .00537 | 3 | .00076 | .00078 | .00078 |
| n=9 | 2.0 | .000018 | .000018 | .000018 | 5 | .000082 | .000088 | .000088 |
| | 2.5 | .000004 | .000005 | .000005 | 7 | .000018 | .000021 | .000021 |
| | 3.5 | .000002 | .000003 | .000003 | 9 | .000006 | .000006 | .000007 |

**Exhibit 6.4** (cf. Daniels, 1983)
Tail probability of Huber's M-estimate with $k = 1.5$

It is helpful in understanding the mechanism of the approximation to consider the case where the $X_i$'s are normal with mean $\mu$ and variance $\sigma^2$. In this case,

$$K(z) = \mu z + \frac{\sigma^2}{2} z^2$$

and $K'(z) = x_0$ gives $\alpha = (x_0 - \mu)/\sigma^2$.
The transformation from $z$ to $t$ is given by $z = t/\sigma$ and the function $G_0(t) = 1$. Hence $a_0 = 1$, $a_1 = 0$ and $H_0(t) \equiv 0$ and the only non-zero term in the approximation is $1 - \Phi(\sqrt{n}\gamma)$ where $\gamma = (x_0 - \mu)/\sigma$ i.e. the approximation (6.3) gives

$$P(\bar{X} \geq x_0) = 1 - \Phi(\sqrt{n}(x_0 - \mu)/\sigma)$$

for normal random variables and is, of course, exact.

In the general case, the function $K(z) - zx_0$ is not parabolic and the Lugannani-Rice approximation proceeds by distorting $K(z) - zx_0$ so that it is parabolic. The first term in the approximation, $1 - \Phi(\sqrt{n}\gamma)$ comes from the normal approximation. The next terms arise from the non-linearity of the transformation of $K(z) - zx_0$ to a parabola. It should be noted that a different normal (or parabolic) approximation is used for each point $x_0$. This approach of a local normal approximation is of course the same as that used in deriving the small sample approximation for the density. The classic Edgeworth and the Fisher-Cornish inversion come from approximating $K(z)$ globally by a polynomial which behaves like $K(z)$ at the origin and matches derivatives at the origin. The local nature of (6.3), ensures the good accuracy of (6.3) even for very small $n$.

We now consider the so-called index of large deviations, namely

$$-\lim_{n\to\infty} \log P(T_n > t)/n$$

where $T_n$ is an estimate. If $T_n$ is a M-estimate of location with a monotone score function, then using a simple Laplacian approximation to the integrated density approximation, we have

$$P(T_n > t) = c^{-n}(t)/[\sigma(t)\alpha(t)(2\pi n)^{1/2}][1 + 0(1/n)].$$

This approximation works as well as (6.3) in the extreme tails. The extra terms in (6.3) are an adjustment for the situation when $t$ is near the singularity in the integral at the mean. From this it follows that

$$-\lim_{n\to\infty} \log P(T_n > t)/n = \log c(t)$$

In addition, if $T_n$ converges almost everywhere to $T(F)$, then the Bahadur half slope is given by

$$\lim_{n\to 0} \frac{1}{n} \log \int_{T_n}^{\infty} f_n(u)du = \lim_{n\to\infty} (-\log c(T_n))$$

$$= -\log c(T(F))$$

(see Bahadur 1971, section 7).

*Remark 6.1*
We can now complete the proof of (4.9). Let $g_n(t) = (n/2\pi)^{1/2}c^{-n}(t)A(t)/\sigma(t)$ denote the approximation to the density. The Laplacian approximation above for $\int_t^\infty g_n(t)dt$ gives that

$$\lim_{n\to\infty} n \int_t^\infty g_n(t)dt = \lim_{n\to\infty} \frac{n^{1/2}\exp(-n\log c(t))}{\sigma(t)\alpha(t)(2\pi)^{1/2}}\left(1+0\left(\frac{1}{n}\right)\right).$$

Since $\log c(t) > 0$ for $t > \theta_0$, it follows directly that the above limit is 0 as required to give $\int_{\delta_2+\theta_0}^\infty g_n(t)dt = 0(1/n)$.

*Remark 6.2*
It is worth noting that in the tail area approximation (6.3), we get very accurate results for the tail area beyond $t$ using only the characteristics of the conjugate density at $t$. This suggests that the conjugate density is the natural mechanism for centering and accurately captures the tail area behavior.

To conclude this section we look at some results of Dinges (1986a,b,c) in which he tries to develop a coherent theory for distributions close to the normal. He obtains an approximation for the tails of various distributions including Student's $t$, Beta, Inverse normal, and Gamma. Moreover, he applies his technique to approximate the tail area of the distribution of the average of $n$ iid random variables and that of M-estimators of location. Here we will discuss the connection between this approximation and those discussed so far.

The key point is the concept of Wiener germ which is defined as follows in Dinges (1986b).

Let $U$ be a neighborhood of $t_0$. A *Wiener germ* of order $m$ on $U$ with center $t_0$ is a family of densities $\{f_\epsilon(t)|\epsilon \to 0\}$ of the following form

$$f_\epsilon(t) = (2\pi\epsilon)^{-p/2}\exp\{-K^*(t)/\epsilon\}\cdot D(t)\cdot\exp\{\epsilon\cdot S(\epsilon,t)\}, \tag{6.4}$$

where

a) $K^*(t)$, the entropy function , is $(m+1)$-times continuously differentiable with $K^*(t_0) = 0$, $K^*(t) \geq 0$, $K^{*\prime\prime}(t)$ positive definite;
b) $D(t)$, the modulating density , is positive and $m-$times continuously differentiable;
c) $\epsilon\cdot S(\epsilon,t) = \epsilon\cdot S_1(t)+\epsilon^2 S_2(t)+\cdots+\epsilon^{m-1}S_{m-1}(t)+\epsilon^m R(\epsilon,t)$, with the correcting functions $S_j(t)$ $(m-j)$-times continuously differentiable and the remainder term $R(\epsilon,t)$ uniformly bounded on $U$;
d)

$$\int_{t'}^{t''} f_\epsilon(t)dt = 1 - 0(\epsilon^m), \tag{6.5}$$

for some $t' < t_0 < t''$, $[t',t''] \subseteq U$.

The following interpretation will help to clarify this notion. Consider a diffeomorphism $V_0 : G \to U \subset \mathbf{R}^p$, where $G$ is a neighborhood of the origin. Let $V_1(\cdot),\cdots,V_m(\cdot)$ be differentiable mappings on $G$. For small $\epsilon$ consider the mappings

$$V(\epsilon,\cdot) = V_0(\cdot) + \epsilon V_1(\cdot) + \cdots + \epsilon^m V_m(\cdot) \tag{6.6}$$

and denote by $W(\epsilon,\cdot) : U \to G_\epsilon$ its inverse. If the $V_j$ are sufficiently smooth, there exist $W_0(t),\cdots,W_m(t)$ such that

$$W(\epsilon,t) = W_0(t) + \epsilon W_1(t) + \cdots + \epsilon^m W_m(t) + 0(\epsilon^{m+1}). \tag{6.7}$$

Then the distribution restricted to $G_\epsilon$ of a normal random variable with expectation 0 and covariance matrix $\epsilon I$ is mapped by $V(\epsilon,\cdot)$ into a distribution on $U$ with a density of the form $f_\epsilon(t)$ given by (6.4). Near the center $t_0$ the densities $f_\epsilon(t)$ are similar to the normal densities with mean $t_0$ and covariance matrix $\epsilon(K^{*''}(t_0))^{-1}$.

An example of Wiener germ is provided by Daniels' result (3.19) written in terms of Legendre transform (see section 5.2). In fact, given a sequence of random variables $X_1,\cdots,X_n$, the densities of the mean $\bar{X}_n = n^{-1}\sum_{i=1}^{n} X_i$ follow a Wiener germ along $\epsilon = \epsilon_n = \frac{1}{n}$. In this case the entropy function $K^*$ is the Legendre transform of the cumulant generating function, i.e. $K^*(t) = \sup\{\tilde{\alpha}t - K(\tilde{\alpha}) : \tilde{\alpha}\} = \alpha K'(\alpha) - K(\alpha)$ with $\alpha = \alpha(t)$ determined by $K'(\alpha) = t$, and the modulating density is $D(t) = (K^{*''}(t))^{1/2}$.

Dinges gives then the following expansion for the tail area of a Wiener germ of order $m$

$$\int_t^\infty f_\epsilon(y)dy = \Phi\Big\{-\epsilon^{1/2}[W_0(t) + \epsilon W_1(t) + \cdots + 0(\epsilon^m)]\Big\}, \tag{6.8}$$

uniformly in the interval $[t_0 - c\sqrt{\epsilon}, t'']$, where $c$ is an arbitrary but fixed positive number and $\Phi$ is the cumulative of the standard normal distribution. The coefficients of the expansion (6.8) $W_0, W_1, \cdots$ have a direct interpretation as the coefficients in the expansion (6.7) of the inverse $W$ of the mapping $V(\epsilon,\cdots)$. They can be computed iteratively from the entropy function $K^*$, the modulating density $D$, and the correcting functions. For instance,

$$W_0(t) = (2K^*(t))^{1/2}$$

$$W_1(t) = -\frac{1}{W_0(t)}\log\left[(K'(t))^{1/2}/W_0'(t)\right].$$

When (6.8) is applied to the distribution of the mean $\bar{X}_n$ of $n$ iid random variables one obtains the following expansion ($\epsilon = \epsilon_n = 1/n$)

$$P[\bar{X}_n > t] = \Phi\Big\{-\sqrt{n}[W_0(t) + \frac{1}{n}W_1(t) + \cdots o(n^{-m})]\Big\}$$

or

$$-\frac{1}{\sqrt{n}}\Phi^{-1}\{P[\bar{X}_n > t]\} = W_0(t) + \frac{1}{n}W_1(t) + \cdots o(n^{-m}), \tag{6.9}$$

where

$$W_0(t) = (2K^*(t))^{1/2} = \alpha\cdot K'(\alpha) - K(\alpha),$$

$$W_1(t) = \frac{1}{W_0(t)}\log[w(t)/W_0(t)],$$

$$w(t) = K^{*'}(t)\cdot(K^{*''}(t))^{1/2},$$

and $\alpha(t)$ is determined by the saddlepoint equation $K'(\alpha) = t$ or equivalently $K^{*'}(t) = \alpha$. In other words, one obtains up to the term of order $1/n$ in (6.9).

$$P[\bar{X}_n > t] \sim \Phi\Big\{-\sqrt{n}W_0(t) - \frac{1}{\sqrt{n}W_0(t)}\log[w(t)/W_0(t)]\Big\}. \tag{6.10}$$

A similar result can be obtained for M-estimators of location.

The expansion (6.9) should be compared with large deviations types of results where the quantity $\frac{1}{n}\log\{P[\bar{X}_n > t]\}$ is expanded. Since

$$\frac{1}{2}\left[\Phi^{-1}(p)\right]^2 \sim \log\frac{1}{p} \quad \text{when} \quad p \to 0,$$

Dinges argues from (6.9) that $-\Phi^{-1}\{P[\bar{X}_n > t]\}/\sqrt{n}$ is the quantity that ought to be expanded. Finally, note that a term expansion of formula (6.10) is similar to Lugannani and Rice (1980) expansion (see 6.3) given by (in this notation)

$$P[\bar{X}_n > t] \sim \Phi(-\sqrt{n}W_0(t)) + \frac{1}{\sqrt{n}}\phi(\sqrt{n}W_0(t))\left[\frac{1}{w(t)} - \frac{1}{W_0(t)}\right]$$

where $\phi$ is the standard normal density.

## 6.3. CONFIDENCE INTERVALS

In many estimation situations, it is of substantial interest to compute confidence intervals for parameters of interest. We have developed techniques for approximating densities in the previous chapters and our aim here is to use those approximations to construct confidence intervals. Consider the setting of section 4.5, namely that of an independent identically distributed sample $X_1, \cdots, X_n$ drawn from a population $F_\eta$ involving an unknown p-dimensional parameter $\eta$. $\eta$ is estimated by an M-estimate $\hat{\eta}$ as the solution of

$$\sum_{i=1}^{n} \psi_j(x_i, \eta) = 0, \quad j = 1, \cdots, p.$$

We want to construct an interval for a real-valued parameter $\theta = \theta(\eta)$. $\theta$ will be estimated by $\hat{\theta} = \theta(\hat{\eta})$. In a linear regression problem $\eta = (\beta_0, \beta_1, \sigma)$ where $\beta_0$ is the intercept and $\beta_1$ the slope and we are often interested in $\theta(\eta) = \beta_1$ as a parameter of interest. In a testing context $\theta$ may be the value of a test statistic. In this case, we are more interested in computing tail areas for $\hat{\theta}$. The confidence intervals will be constructed via test statistics and tail areas so that if the interest is in P-values, they can easily be obtained.

If we start with approximation (4.25) for $f_n(\mathbf{t})$, the density of $\hat{\eta}$, there are several problems arising in the construction of confidence intervals. The first is the choice of an appropriate test procedure to see whether a trial value $\theta_0$ belongs to the interval or not. Given an appropriate test statistic, we need to compute a P-value on which to base our decision about $\theta_0$. Although we have an approximation for the density of $\hat{\eta}$, to compute the marginal density of $\hat{\theta}$ requires evaluation of $f_n$ over a p-dimensional space and then effectively integrating out $(p-1)$ dimensions. Each evaluation of $f_n$ requires the solution of a system of $p$ nonlinear equations. This procedure is feasible for $p = 2$ but becomes computationally infeasible for larger $p$'s . We need to find a technique to reduce the computational complexity. Even without the computational difficulties, there still remains the problem of how to handle the nuisance parameters in constructing the interval for $\theta$. As a final difficulty we may not want to specify $F_\eta$. The more natural way to overcome this is to replace $F_\eta$ by an appropriate empirical density. This means that we are using a philosophy similar to that of the bootstrap but as we shall show, we avoid the resampling associated with the bootstrap. The procedure outlined here is based on Tingley and Field (1990).

To begin, we assume the assumptions in section 4.5 are met. Then from (4.24) we can write

$$(\hat{\eta} - \eta_0)_r = \sum_{j=1}^{p} b_{rj} \bar{Z}_j + o_p(1/\sqrt{n}) \tag{6.11}$$

where $\bar{Z}_j = \sum_{i=1}^{n} \psi_j(X_i, \eta_0)/n$ and $B = (b_{rj}) = -A(\eta_0)^{-1}$ and $A(\eta_0) = E_{\eta_0}\left[\frac{\partial \psi}{\partial \eta}(X, \eta_0)\right]$.

We now want to consider testing a point $\theta_0$ to see whether it belongs to the interval or not. $\eta_0$ is unknown but satisfies $\theta(\eta_0) = \theta_0$. To construct a test statistic we expand $\hat{\theta} = \hat{\theta}(\eta)$ in a Taylor series expansion about $\eta_0$ using (6.11) and obtain

$$\hat{\theta} - \theta(\eta_0) = \sum_{i=1}^{n} g(X_i, \eta_0)/n + 0_p(1/n)$$

where $g(X_i, \eta_0) = \psi^T(X_i, \eta_0) B^T \partial\theta(\eta_0)/\partial\eta$. The random variables $g_i = g(X_i, \eta_0)$ are referred to as the configuration.

Since $\eta_0$ is assumed unknown, we actually work with the observed configuration $g_i = g(X_i, \hat{\eta})$. In order to keep the ideas clear we denote the observed $\hat{\eta}$ by $\hat{\eta}_{obs}$. Both $\hat{\eta}_{obs}$ and $\hat{\theta} = \theta(\hat{\eta}_{obs})$ are held fixed in what follows. Note that $\bar{g}_{obs} = 0$. Our test statistic is $\bar{g}$ which approximates $\theta - \theta_0$ with error $0_p(1/n)$. Although this may not appear to be accurate enough (cf. Hall, 1988), we are able to obtain intervals with good coverage by carefully approximating the density of $\bar{g}$. Since our test statistic is a mean, the small sample approximation for the density of a mean can be used. Before giving the coverage properties of our interval, we give the algorithm to construct the interval and illustrate it with an example.

Step 1:

Compute $\hat{\eta}_{obs}$, $\theta$ and $g_i = \psi^T(x_i, \hat{\eta}_{obs}) B^T \partial\theta(\eta)/\partial\eta$. $\hat{B}$ can be computed parametrically as the inverse of $E_{\hat{\eta}_{obs}}\left[\frac{\partial \psi}{\partial \eta}(X, \hat{\eta}_{obs})\right]$ or nonparametrically as the inverse of $\hat{A} = \sum_{i=1}^{n} \frac{\partial \psi}{\partial \eta}(x_i, \eta)/n$.

Step 2:

Obtain an initial estimate of the distribution of the $g_i$. At this point either a parametric or nonparametric estimate can be used.

In this development we use a nonparametric estimate via the cumulant generating function $K(t)$.

Let $K(t) = \log\left(\sum_{i=1}^{n} e^{tg_i}/n\right)$.

Steps 1 and 2 must be computed once for every sample. Note that the mean of our estimated distribution, $K'(0) = \sum_{i=1}^{n} g_i/n = 0$. For each $\theta_0$ under test we need to recenter the distribution at $\theta_0 - \hat{\theta}$, which is the approximate expected value of $g(Y, \hat{\eta}_{obs})$ under $F_{\eta_0}$. This is accomplished as in previous chapters by solving for $\alpha$ in the equation

$$\sum_{i=1}^{n} (g_i - (\theta_0 - \hat{\theta})) \exp\{\alpha(g_i - (\theta_0 - \hat{\theta}))\} = 0. \tag{6.12}$$

Now (6.12) is the centering result used before except that the density $f$ has been replaced by the empirical distribution function.

Step 3:

For each $\theta_0$ under test, compute $\alpha(\theta_0)$ as the solution of

$$\sum_{i=1}^{n}(g_i - (\theta_0 - \hat{\theta})(1-e))\exp\{\alpha(g_i - (\theta_0 - \hat{\theta})(1-e))\} = 0$$

where $e$ is of order $1/n$.

The correction term $e$ will be discussed later and is effectively a calibration correction.

Step 4:

Approximate $P_{\eta_0}[\bar{g} > 0]$ by

$$\hat{P}_{\theta_0}[\bar{g} > 0] = \Phi(-\sqrt{2n_1 \log c(\theta_0)}) - \frac{c^{-n_1}(\theta_0)}{\sqrt{2\pi n_1}}[\frac{1}{a(\theta_0)\sigma(0)} - \frac{1}{\sqrt{2\log c(\theta_0)}}]$$

i.e. use the tail area approximation (6.3), where $n_1 = n - p$ or $n - (p-1)$. Include $\theta_0$ in the $(1-2\epsilon)100\%$ interval if $\epsilon < \hat{P}_\theta[\bar{g} > 0] < 1 - \epsilon$. Note that (6.3) is used with cumulant generating function $K(t) = K(t + \alpha_0) - K(t)$ where $\alpha_0 = \alpha(\theta_0)$.

It is more efficient to work backwards and replace steps 3 and 4 by

Step 3′ :

Find $\alpha_1$ solutions of $P[\bar{g} > 0] = \epsilon$ and $P[\bar{g} > 0] = 1 - \epsilon$ where

$$\hat{P}[\bar{g} > 0] = \Phi(-\sqrt{2n_1 \log c(\alpha)} - \frac{c^{-n_1}(\alpha)}{\sqrt{2\pi n_1}}\left\{\frac{1}{\alpha\sigma(0)} - \frac{1}{\sqrt{2n_1 \log c(\alpha)}}\right\}.$$

We write $c$ and $s$ as depending on $\alpha$ in this case, rather than $\theta_0$. It is still the integrating constant of the conjugate density.

Step 4′ :

Find $\theta_1$ and $\theta_2$ solutions of

$$\sum_{i=1}^{n}(g_i - (\theta_j - \hat{\theta})(1-e))\exp\{\alpha_j(g_i - (\theta_j - \hat{\theta})(1-e))\} = 0, \quad j = 1, 2.$$

The estimated $(1-2\epsilon)100\%$ confidence interval is $(\theta_1, \theta_2)$.

We now consider using this algorithm for the location/scale problem. For the sample $X_1, \cdots, X_n$ from an unknown distribution, location $\mu$ and scale $\sigma$ are estimated by the M-estimates $\hat{\mu}$, $\hat{\sigma}$, solutions of

$$\frac{1}{n}\sum_{i=1}^{n}\psi(X_i, \eta) = 0$$

where $\eta = (\mu, \sigma)$ and

$$\psi_1(x, \eta) = \psi_k\left(\frac{x-\mu}{\sigma}\right)$$

$$\psi_2(x, \eta) = \psi_k^2\left(\frac{x-\mu}{\sigma}\right) - \beta,$$

$$\psi_k(y) = \begin{cases} -k & \text{if } y \le -k \\ y & \text{if } |y| < k \\ k & \text{if } y \ge k, \end{cases}$$

$$\beta = (n-1)\int_{-\infty}^{\infty} \psi_k^2(y)\phi(y)dy/n,$$

and $\phi(y)$, $\Phi(y)$ are the density function and cumulative distribution, respectively, of a standard normal random variable. The estimate $\eta$ is referred to in the literature as Huber's Proposal 2 (1981). The constant $k$ is usually between 1 and 2. We have used, exclusively, $k = 1.0$.

If the matrix $A = E[\partial\psi/\partial\eta]$ is calculated under the normal model, then

$$A = -\frac{1}{\sigma}\begin{bmatrix} \delta & 0 \\ 0 & 2\epsilon \end{bmatrix}$$

where $\delta = \Phi(k) - \Phi(-k)$ and

$$\epsilon = \int_{-k}^{k} y^2\phi(t)dy,$$

Then $B = -A^{-1}$ is estimated by

$$\hat{B} = \hat{\sigma}\begin{bmatrix} 1/\delta & 0 \\ 0 & 1/2\epsilon \end{bmatrix}$$

and

$$g_i = \psi^T(x_i, \hat{\eta})\hat{B}^T\partial\theta(\hat{\eta})/\partial\eta = \hat{\sigma}\psi_k((x_i - \hat{\mu})/\hat{s}\Big)/\delta. \tag{6.13}$$

If the matrix $A$ is estimated empirically, then

$$g_i = \frac{\hat{\sigma}}{a_{11}a_{22} - a_{12}^2}\left\{a_{22}\psi_k\left(\frac{x_i - \hat{\mu}}{\hat{\sigma}}\right) - \frac{a_{12}}{2}\left(\psi_k^2\left(\frac{x_i - \hat{\mu}}{\hat{\sigma}}\right) - \beta\right)\right\}. \tag{6.14}$$

where the $a_{ij}$ are the elements of $\hat{A}$ (see step 1).

By looking at numerical results, it can be shown that (6.14) is non-robust in that a small shift in an observation can lead to a large change in the configuration. On that basis we recommend the use of (6.13).

The following table taken from Tingley and Field (1990) shows the results which are obtained using a Monte Carlo swindle and were blocked by generating 2000 samples of size 30 and then using subsets of these for samples of size 5, 10 and 20. The classical samples are the usual t-intervals while the small sample intervals are as given above with $k = 1$.

| Sample Size | | Normal Classical | Normal Small sample | Contaminated normal .9N(0,1) + .1N(0,16) Classical | Contaminated normal Small sample | $t_3$ Classical | $t_3$ Small sample |
|---|---|---|---|---|---|---|---|
| | 2 sided | | | | | | |
| | coverage | .95 | .903 | .960 | .916 | .963 | .921 |
| 5 | $P < R$ | .975 | .950 | .980 | .958 | .98 | .96 |
| | s.d | - | .07 | .010 | .063 | .02 | .06 |
| | log length | .98 | .91 | 1.30 | 1.07 | 1.38 | .15 |
| | 2 sided | | | | | | |
| | coverage | .95 | .929 | .961 | .929 | .960 | .931 |
| 10 | $P < R$ | .975 | .960 | .981 | .965 | .98 | .96 |
| | s.d | - | .05 | .015 | .053 | .03 | .06 |
| | log length | .39 | .41 | .73 | .51 | .79 | .59 |
| | 2 sided | | | | | | |
| | coverage | .95 | .937 | .960 | .939 | .955 | .937 |
| 15 | $P < R$ | .975 | .97 | .981 | .971 | .98 | .96 |
| | s.d | - | .04 | .024 | .041 | .06 | .06 |
| | log length | -.05 | -.02 | .32 | .08 | .38 | .14 |
| | 2 sided | | | | | | |
| | coverage | .95 | .943 | .958 | .941 | .954 | .939 |
| 20 | $P < R$ | .975 | .97 | .979 | .971 | .98 | .96 |
| | s.d | - | .03 | .033 | .037 | .06 | .06 |
| | log length | -.34 | -.29 | .05 | -.20 | .11 | -.15 |

**Exhibit 6.5**

95% Confidence intervals for location

The results from the table show that the coverage is always slightly less than the required level of .95. This can be adjusted by fine-tuning the effective degrees of freedom or adjusting the shift condition. Work is continuing on the geometric interpretation of these adjustments. The results on the lengths show that for both the contaminated normal and $t_3$, the length of the classical interval is about 25% longer than that of the small sample interval. Results for the correlation coefficient and the percentiles of a $t_3$-density (first and tenth) are reported in Tingley and Field (1990). In both instances, the small sample intervals give actual coverage very close to the desired level. In particular for the correlation coefficient, the coverage is always better and the length shorter than that of the Fisher interval.

We now turn to a discussion of the procedure and its coverage. As can be noted from the algorithm, the interval is constructed by inverting the test $H_0 : \theta(\eta_0) = \theta_0$ where $\eta_0$ is unknown. Recall that in this argument, $\hat{\eta}_{obs}$ and $\hat{\theta} = \theta(\hat{\eta}_{obs})$ are being held fixed. With $\eta_0$, it can be verified that $E(g(X, \hat{\eta}_{obs})) = \theta_0 - \hat{\theta}_{obs}$ with error $O_p(1/n^{3/2})$. The approximation to the density of $g$ under $\eta_0$ is given by the cumulant generating function $K(t)$ defined by

$$K(t) = \log\left(\frac{1}{n}\sum_{i=1}^{n} \exp(tg(x_i, \hat{\eta}_{obs}))\right).$$

However with $K(t)$, we have that the mean of $g(X, \hat{\eta}_{obs})$ is 0 while in fact the mean should be $\theta_0 - \hat{\theta}$ . This suggests that $K(t)$ is not a good estimate of the density. The centering conditon (6.12) modifies our estimate $K(t)$ to $K_{\alpha_0}(t) = K(\alpha_0+t) - K(\alpha_0)$. The calculations involving $g$ under $\eta_0$ with $\theta(\eta_0) = \theta_0$ are now done with the approximate density determined by $K_{\alpha_0}(t)$. Consider the one-dimensional exponential family $h^*(g;\alpha) = \exp(\alpha g - K(\alpha))h(g,\eta)$ where $h(g,\eta)$ is the density of $g$ determined by $K$. Then the hypothesis $H_0 : \theta(\eta_0) = \theta_0$ is equivalent (to order $0_p(1/n^{3/2})$) to the hypothesis $H_0 : \alpha = \alpha_0$ in the exponential family $h^*$. For such an exponential family, we know that the uniformly most powerful test of $H_0 : \alpha = \alpha_0$ versus $H_1 : \alpha > \alpha_0$ is to reject $H_0$ if $\bar{g}_{obs}$ exceeds a critical value. Our confidence intervals are now constructed by computing $P_{\theta_0}[\bar{g} > 0]$ using the approximation in step 4 and inverting the test. Conditional on the density determined by $K(t)$ and the assumed one-parameter exponential family $h^*(g;\alpha)$, the confidence intervals constructed above are optimal.

The final step is to obtain results on the coverage. A straightforward application of an Edgeworth expansion gives bounds on the error of the cumulants from using $K_{\alpha_0}(t)$ and we obtain

$$P_{\eta_0}[\bar{g} < 0] = \hat{P}_{\theta_0}[\bar{g} < 0] + 0_p(1/n) \quad \text{if} \quad \theta(\eta_0) = \theta_0. \tag{6.15}$$

Expression (6.15) guarantees that the constructed intervals are second order correct for one-sided coverage. See Tingley and Field (1990) for a proof of (6.15) along with further discussion. Hall (1988) gives an excellent exposition on bootstrap confidence intervals.

The importance of the algorithm is that it resolves two problems with the usefulness of the small sample approximation in some practical settings. The first is that of the computational complexity in multiparameter problems. The use of the configuration reduces the multiparameter problem to a one-dimensional problem. Secondly by using $K(t)$ and $K_\alpha(t))$ we avoid the problem of specifying an underlying density and allow the data to determine the interval directly. In this sense, the procedure is closely analogous to the bootstrap.

Finally we note that there is no inherent difficulty in handling multiple regression, logistic regression or any procedure where the estimate is defined via a solution of a system of equations. Research is currently underway in applying the technique to a variety of situations.

# 7. FURTHER APPLICATIONS

## 7.1. COMPUTATIONAL ASPECTS

To evaluate the small sample approximations developed in the previous chapters we require the use of numerical software. The most critical aspect involves the solution for the saddlepoint, $\alpha(\mathbf{t})$. In the most general case, this involves solving the system

$$\int \psi_j(x,\mathbf{t}) \exp\left\{\sum_{i=1}^{p} \alpha_j(\mathbf{t})\psi_j(x,\mathbf{t})\right\} f(x)dx = 0 \quad j = 1,\cdots,p. \tag{7.1}$$

This is a non-linear system of equations which require an integration to evaluate the left hand side. Given the values for $\alpha$, the computation of $c(\mathbf{t})$, $\Sigma(\mathbf{t})$, $A(\mathbf{t})$ also involve the evaluation of integrals. In all the examples for which we have done computations, $x$ has been univariate. In principle, there is no difficulty with multivariate $x$, but the problems of integration become severe as the dimension of $x$ increases.

In terms of computational effort, it is the solution of (7.1) which is the dominant feature. In the problems involving location/scale, the functions involved in (7.1) are continuous and piecewise differentiable and the resulting $\alpha_j(\mathbf{t})$'s have been smooth. This smoothness is very helpful in solving (7.1) over a grid of $\mathbf{t}$ since we are able to get good initial guesses for $\alpha$ based on values at adjacent grid points. For one dimensional problems, we have used the NAG (Numerical Algorithm Group) subroutine CØ5AJF. This procedure works well given a good initial guess and iterates using a secant method. The algorithm solves a sequence of problems $h(\alpha) - \theta_r h(\alpha_0)$ where $1 = \theta_0 > \theta_1 > \theta_2 \cdots > \theta_m = 0$ where $\alpha_0$ is the initial guess. For each $\theta_r$, a robust secant iteration is used based on the solution from earlier problems. If bounds are available on the solution $\alpha$, a routine such as CØ5ADF may be used. In the case of location/scale problems $\alpha$ is two-dimensional. For the problem of Huber's Proposal 2, we used the NAG subroutines CØ5NBF or CØ5NCF. Both routines are suitable for the situation where the derivatives with respect to $\alpha$ are not provided. These routines are similar to the IMSL routines HYBRD and HYBRD1. It is necessary to choose initial estimates carefully using information from other grid points. It is possible to evaluate the derivatives of $\alpha$ and use a more reliable routine. However the derivatives are somewhat complicated and it is not clear that the extra coding required is worth the effort.

The numerical integration has been carried out using Gaussian quadrature. The particular NAG subroutine used is DØ1FBF. From experience, it appears that a 32 point quadrature procedure is necessary to get reliable results in (7.1) for $\mathbf{t}$ in the tail of the distribution.

The evaluation of tail areas has been carried out using the Lugannani-Rice approximation (6.3). The error in that approximation is about the same order as that resulting from a numerical integration of $f_n(\mathbf{t})$.

To give a sense of the computing time involved, we did several runs on a SUN4, 0S (4.2 Berkeley Unix) using the NAG subroutines mentioned above. The first two cases involved computation of $\alpha(t)$, $c(t)$, $s(t)$ for the mean, $\psi(x,t) = (x-t)$ for a uniform density on $[-1,1]$ and an extreme density. The results are as follows.

| | | |
|---|---|---|
| Uniform: | $t = 0[.005].99$ | CPU time = 2.5 seconds |
| Extreme: | $t = -9[.1]2.5$ | CPU time = 5.9 seconds |

In a two dimensional problem, we solved for $\alpha_1(t)$ and $\alpha_2(t)$ in the case of the extreme density. For a total of 110 grid points ($t_1$ varying, $t_2$ fixed), the CPU time was 8.2 seconds.

The root finding procedure seems to fail if the initial values of a are not close to the final solution.

Our only attempt to find $\alpha$ in a three-dimensional problem was not successful. The root finding subroutine was unable to find a solution within a reasonable numbers of function evaluations. It is likely that tailor-made root finders would have to be constructed for these higher dimension problems.

It is possible to write equation (7.1) as a differential equation and then use a differential equation solver to obtain $\alpha(t)$. For the case $p = 1$, we can write (7.1) as

$$\int (\psi'(x,t) + \alpha' \psi^2(x,t) + \alpha(t)\psi(x,t)\psi'(x,t)) \exp\{\alpha(t)\psi(x,t)\} f(x) dx = 0$$

where $'$ represents differentiation with respect to $t$. Viewing this as a differential equation in $\alpha(t)$, we are able to solve for $\alpha(t)$. Some preliminary runs gave results similar in terms of accuracy and time to those obtained via the rootfinder. This approach seems worth pursuing in higher dimensional problems.

## 7.2. EMPIRICAL SMALL SAMPLE ASYMPTOTICS AND DENSITY ESTIMATION

The small sample asymptotic approximation of the density or the tail area of some statistic developed in chapters 3 through 5 requires the knowledge of the underlying distribution of the observations. From (4.25), for instance, we see that the underlying distribution $F$ enters in the approximation only through the integrals defining $C$, $A$, $\Sigma$ and $\alpha$. To make the technique nonparametric, it is natural to consider replacing $F$ by the empirical distribution function: we obtain then an *empirical small sample asymptotic approximation*. Notice that in this approximation the integrals which appear for instance in (4.25) become sums. This fact greatly reduces the complexity of the computations.

Feuerverger (1989) studied this approximation for the univariate mean and showed that an appropriate standardization of the estimator allows to keep a relative error of order $O_p(n^{-1/2})$. Ronchetti and Welsh (1990) extended this result to multivariate M-estimators and proved that by appropriate standardization of the estimator the renormalized empirical approximation has a relative error or order $O_p(n^{-1})$. In section 6.3 a similar idea was used to construct confidence intervals based on the empirical cumulant generating function. In that situation, the exponential tilt was used to adjust the first cumulant enabling us to obtain second order coverage for the confidence interval. To summarize: the empirical small sample asymptotic approxiamtion is an alternative to the bootstrap with the advantage that resampling is avoided. For a related result, see Davison and Hinkley (1988).

A related potential application is the use of small sample asymptotics in density estimation. The approximations which have been developed are asymptotic expansions in terms of $n$. Since they work so well for very small sample sizes, we can examine them with $n = 1$ although clearly it does not make sense to discuss the order of the error terms. The approximation for $n = 1$ can be thought of as a version of the underlying density smoothed in some sense towards normality.

Consider the small sample asymptotic approximation of univariate M-estimators given by (4.8). For $n = 1$ we obtain

$$g_1(t) \propto A(t)/(c(t)\sigma(t)),$$

where $A$, $c$, and $\sigma$ are defined in Theorem 4.3. $g_1(t)$ depends on the function $\psi$ which defines the M-estimator. Let us consider this approximation for the mean ($\psi(x,t) = x - t$) and the following four underlying situations:

1. Uniform on $[-1, 1]$
2. Extreme density, $f(x) = \exp(x - \exp(x))$
3. Cosine density on $[0, 1]$, $f(x) = 1 + \cos 4\pi x$
4. U-shape density on $[-1, 1]$, $f(x) = 1.5x^2$

The last two densities were chosen as situations where the density is very different from a normal and we might expect the approximation to do badly. For each of the above, we have calculated $g_1$ and plotted it with the true underlying density $f$ in Exhibit 7.1. As can be seen from the plots, $g_1$ is a reasonable estimate of the density for the extreme but as expected is a poor estimate for the cosine and U-shape although it starts to match reasonably well in the tails. Exhibits 3.9 through 3.13 confirm this result.

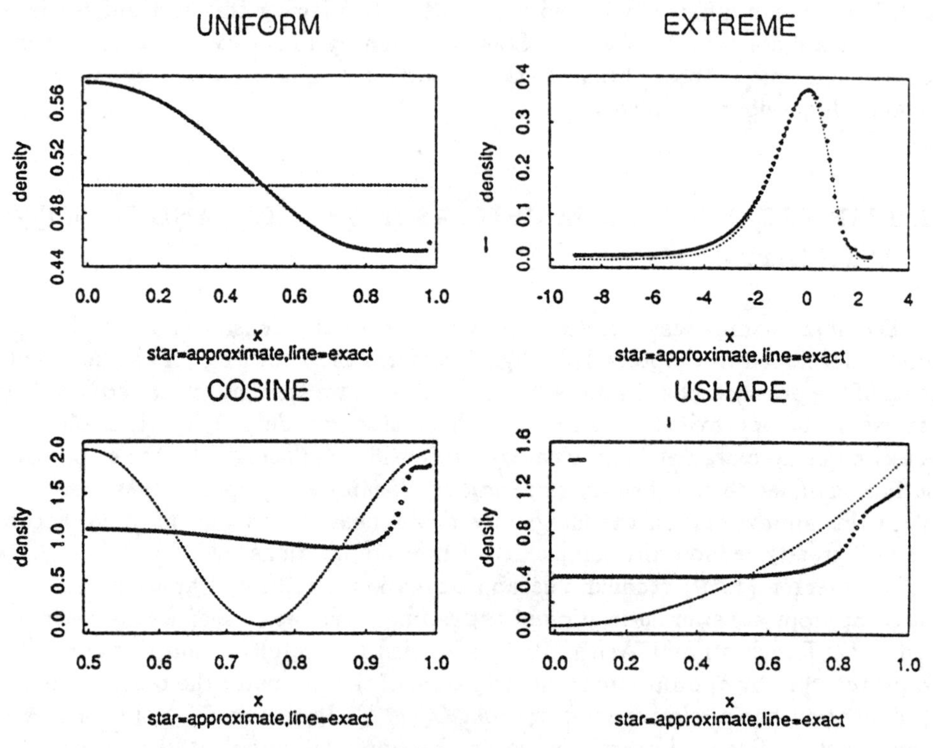


**Exhibit 7.1**
Small sample approximation with $n = 1$ versus true density

If we want to use $g_1$ as a density estimator, we face two problems. Firstly, we have different possible choices of $\psi$. Each of them will define a different estimator. Secondly, the underlying density (which we are trying to estimate) enters in the computation of $g_1$ through $A$, $c$, and $\sigma$.

The second problem can be solved by replacing the underlying distribution in $A$, $c$, and $\sigma$ by the empirical distribution function. If we apply this idea to $g_1(t)$ we obtain the following *nonparametric density estimator* $\hat{g}_1(t)$ for $f(t)$ (see Ronchetti, 1989):

$$\hat{g}_1(t) = \hat{D}_1\hat{A}(t)/(\hat{c}(t)\hat{\sigma}(t)), \tag{7.2}$$

where $\hat{\alpha}(t)$ is determined by the implicit equation

$$\sum_{i=1}^{n} \psi(x_i, t) \exp\{\hat{\alpha}(t)\psi(x_i, t)\} = 0,$$

and

$$\hat{c}(t) = \left( \frac{1}{n} \sum_{i=1}^{n} \exp\{\hat{\alpha}(t)\psi(x_i, t)\} \right)^{-1},$$

$$\hat{A}(t) = \hat{c}(t) \frac{1}{n} \sum_{i=1}^{n} \frac{\partial \psi(x_i, t)}{\partial t} \exp\{\hat{\alpha}(t)\psi(x_i, t)\},$$

$$\hat{\sigma}^2(t) = \hat{c}(t) \frac{1}{n} \sum_{i=1}^{n} \psi^2(x_i, t) \exp\{\hat{\alpha}(t)\psi(x_i, t)\},$$

$$\hat{D}_1 = \left( \int A(t)/(\hat{c}(t)\hat{\sigma}(t)) dt \right)^{-1}.$$

We can investigate the quality of this estimator by looking at $\hat{g}_1(t) - f(t)$. This difference can be written as

$$\hat{g}_1(t) - f(t) = [\hat{g}_1(t) - g_1(t)] + [g_1(t) - f(t)].$$

Whereas the first term (variability) decreases as $n$ increases, the second one (bias) is fixed. Clearly this term depends on the choice of $\psi$ and on the underlying density $f$ and plays an important role in determining the quality of the estimator.

Feuerverger (1989) studied this problem in the case of the mean ($\psi(x, t) = x - t$) and showed that the bias can be substantial; see Exhibit 7.2.

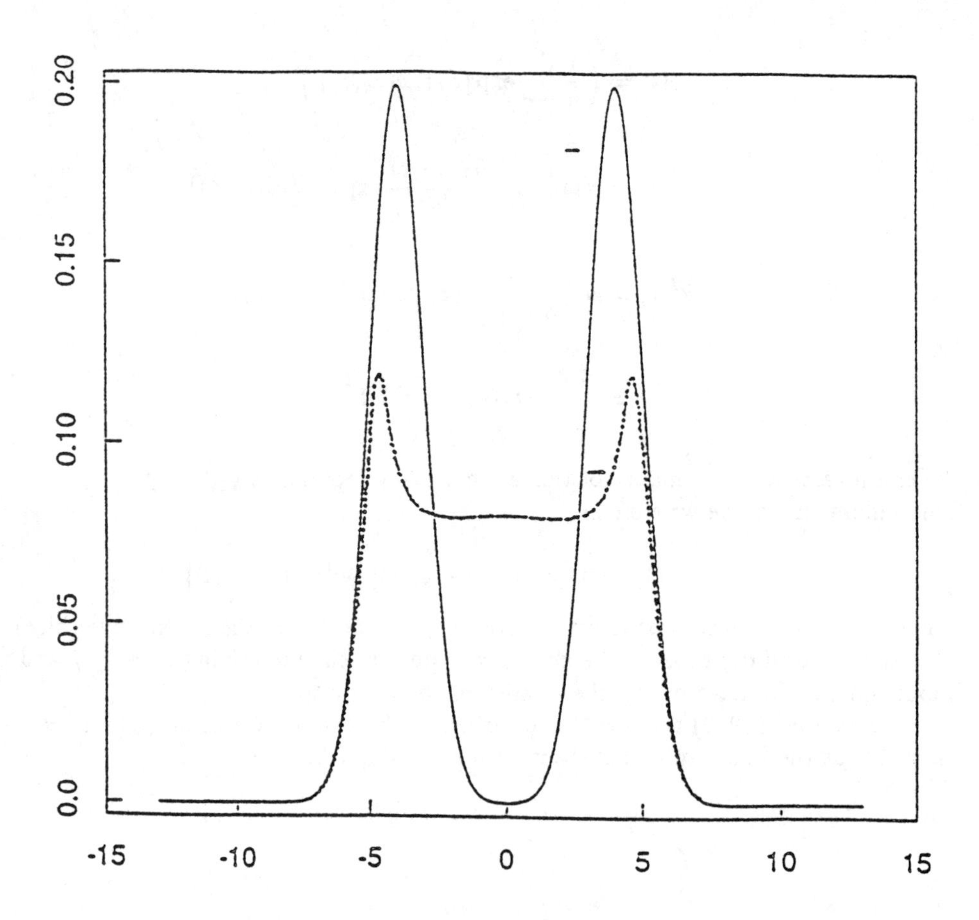


**Exhibit 7.2**
$f(t)$ and $g_1(t)$ (- - -) for the underlying distribution
$.5^*N(4,1) + .5^*N(-4,1)$; from Feuerverger (1989).

Ronchetti (1989) proposed an estimator based on the following $\psi$ function:

$$\begin{aligned} \psi_{k_n}(y) &= y/k_n \qquad &\text{if } |y| \le k_n \\ &= sgn(y) \qquad &\text{otherwise,} \end{aligned}$$

where $k_n \to 0$ as $n \to \infty$. With this choice of $\psi$, (7.2) takes the form:

$$\hat{g}_1(t) = \frac{1}{n}\sum_{i=1}^{n} \frac{1}{2k_n} 1\left\{ \frac{|x_i - t|}{k_n} \le 1 \right\} \exp\{\alpha_n(t)\psi_{k_n}(x_i - t)\}.$$

$$\left[ \frac{\sum_j \exp\{\alpha_n(t)\psi_{k_n}(x_j - t)\}}{\sum_j \psi_{k_n}^2(x_j - t)\exp\{\alpha_n(t)\psi_{k_n}(x_j - t)\}} \right]^{1/2} . \qquad (7.2a)$$

where $1\{\ \}$ is the indicator function and $\alpha_n(t)$ is defined through the implicit equation

$$\sum_{i=1}^{n} \psi_{k_n}(x_i - t)\exp\{\alpha_n(t)\psi_{k_n}(x_i - t)\} = 0.$$

Note that $\hat{D}_1$ in (7.2) is replaced here by its asymptotic value 1/2. The reason for this choice of $\psi$ is that $\psi_{k_n}$ converges to the $\psi$-function of the median as $n \to \infty$ and the small sample asymptotic approximation for the median is exact (after renormalization) for any underlying density $f$; see Field and Hampel (1982) and section 4.2. Therefore the bias term of the density estimator (7.2$a$) vanishes asymptotically. A comparison of Exhibits 7.2 and 7.3 shows the important reduction of bias achieved by using the estimator (7.2$a$). More details are provided in Ronchetti (1989).

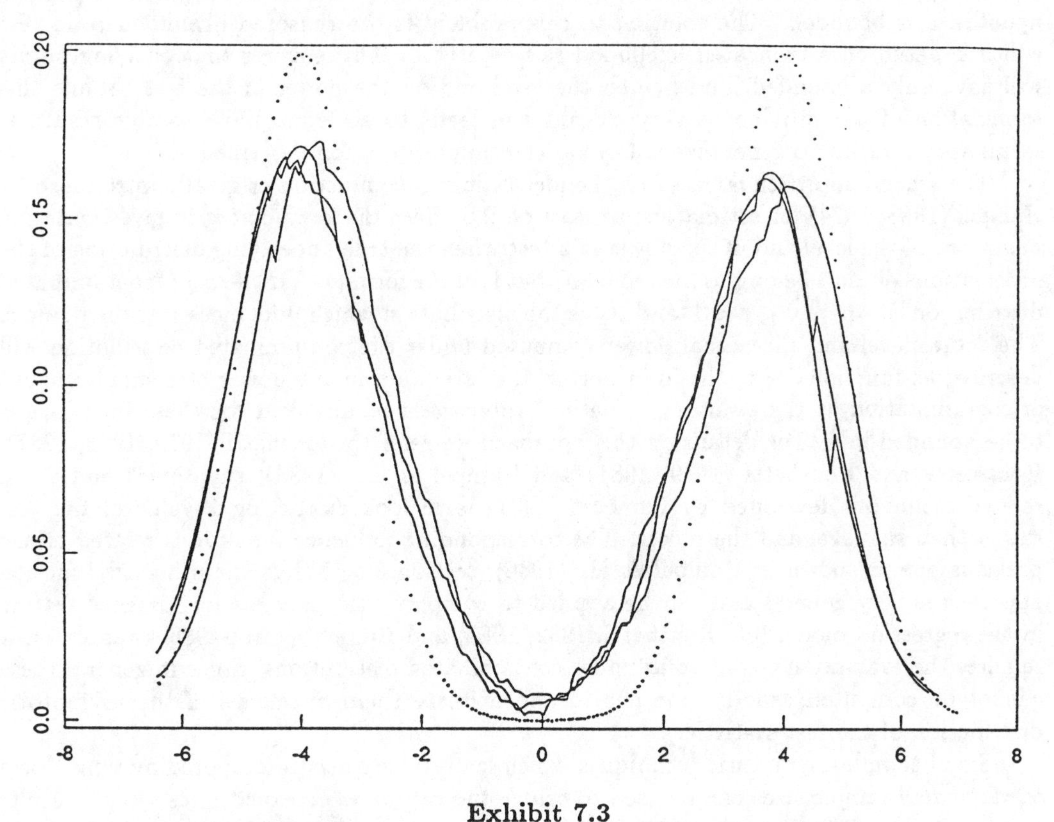


**Exhibit 7.3**
Exact density (- - -) and density estimator for three samples generated from a $.5^*N(4,1) + .5^*N(-4,1)$ and $n = 200$.

Finally it is possible to use $\hat{g}_1(t)$ along with the Lugannani-Rice tail area approximation to construct prediction intervals. Field and Manchester are currently carrying out some research in this direction.

## 7.3. ROBUST TESTING

### 7.3.a The Use of Small Sample Asymptotics in Robust Statistics

In this section we will discuss an application of small sample asymptotics to robust statistics. It shows that small sample asymptotics can be applied successfully not only to compute very accurate approximations to the exact distribution of robust estimators and test statistics (cf. sections 4.2 and 4.5 and Visek, 1983, 1986) but also, from a more methodological point of view, to define new tools and to improve those based on asymptotic normality.

Consider, for instance, the study of the robustness properties of tests. Two different approaches are available.

The first one is Huber's minimax approach (see Huber 1965; 1981, p. 253 ff.) which is based on the following idea. In a simple hypothesis against a simple alternative testing problem, find the test which maximizes the minimum power over a neighborhood of the alternative, under the side condition that the maximum level over a neighborhood of the hypothesis is bounded. The solution to this problem is the censored likelihood ratio test which is based on a truncated likelihood ratio statistic. This ensures that outlying points will have only a bounded influence on the level and on the power of the test. While this formulation of the problem is very elegant and leads to an exact finite sample result, it seems very difficult to generalize it beyond the simple situation described above.

The second approach is based on the idea of influence function originally introduced by Hampel (1968, 1974) for estimators; see section 2.5. Here the key point is to investigate the behaviour of the level and of the power of a test when the true underlying distribution of the observations doesn't belong to the model $F_\theta$ but is of the form $(1-\epsilon)F_\theta+\epsilon\Delta_x$ ("contaminated distribution"), where $0 < \epsilon < 1$, and $\Delta_x$ is the distribution which puts mass 1 at the point $x$. The actual level and the actual power computed under the contaminated distribution will describe, as functions of $x$, the influence on the level and on the power of a small amount of contamination at the point $x$. A natural robustness requirement for these functions is to be bounded in $x$. For details on this approach we refer to Ronchetti (1979, 1982, 1987), Rousseeuw and Ronchetti (1979, 1981) and Hampel et al. (1986), chapters 3 and 7. A related technique developed by Lambert (1981) is to look at the log P-value of the test rather than the level and the power. The corresponding influence function is related to the previous one as shown in Hampel et al. (1986), section 3.6. While the influence function approach is very general and can be applied to complex situations, as for instance testing in the regression model (cf. Ronchetti, 1982, 1987, and Hampel et al., 1986, chapter 7), it requires the evaluation of tail areas under contaminated distibutions. Since in general these cannot be computed exactly, one has to approximate them by means of the asymptotic distribution of the test statistic.

Small sample asymptotic techniques which lead to very accurate approximations down to very small sample sizes can be used to bridge the gap between exact finite sample results and results based on the asymptotic distribution. In sections 7.3b and 7.3c we will illustrate this point for a simple class of tests. The arguments hold for more general situations.

### 7.3.b A Tail Area Influence Function

Suppose we want to investigate the properties of the tail area $P_F(T_n > t)$ ($t$ and $n$ fixed) from a robustness point of view or, more precisely, we are interested in the behavior of $P_F(T_n > t)$ when the observations do not follow exactly the model distribution $F$. To accomplish this, we use the idea of an influence function and introduce a tail area influence function.

**Definition 7.1** Let $\Delta_x$ be the distribution which puts mass 1 at any point $x \in \mathbf{R}$ and define $F_{\epsilon,x} := (1-\epsilon)F + \epsilon\Delta_x$. Then, the *tail area influence function* of $T_n$ at $F$ is defined by

$$TAIF(x;t;T_n,F)=\lim_{\epsilon\to 0}\left[P_{F_{\epsilon,x}}(T_n>t)-P_F(T_n>t)\right]/\epsilon,$$

for all $x\in\mathbf{R}$ where the right hand side exists.

$TAIF(x;t;T_n,F)$ describes the normalized influence on the tail area of a small amount of contamination at the point $x$.

Let us now apply this definition to the following one-parameter situation. Consider a family of distributions $\{F_\theta|\theta\in\Theta\}$ where $\Theta$ is a real interval and a sequence of statistics $T_n(x_1,\cdots,x_n)$ which are used to test the null hypothesis $H_0:\theta=\theta_0$. The test is assumed to be based on a M-statistic, that is the test statistic $T_n$ is the solution of the equation

$$\sum_{i=1}^{n}\psi(x_i,T_n)=0, \tag{7.3}$$

where $x_1,\cdots,x_n$ are $n$ independent observations and $\psi$ is a given function; cf. (4.1).

A small sample asymptotic approximation to the exact density $f_n(t)$ can be obtained from Theorem 4.3. The corresponding approximation to the tail area $P_F(T_n>t)=\int_t^\infty f_n(u)du$ can be computed by means of a Laplacian approximation to the integral when the approximation (4.8) is substituted for $f_n(u)$; cf. section 6.2. This leads to the following tail area approximation

$$P_F(T_n>t)=(2\pi n)^{-1/2}c_F^{-n}(t)/\left[\sigma_F(t)\cdot\alpha_F(t)\right]\cdot\left[1+0(1/n)\right], \tag{7.4}$$

where $\alpha_F(t)$ solves

$$\int\psi(x,t)\exp\{\alpha_F\cdot\psi(x,t)\}dF(x)=0, \tag{7.5}$$

$$c_F^{-1}(t)=\int\exp\{\alpha_F\cdot\psi(x,t)\}dF(x),$$

$$\sigma_F^2(t)=c_F(t)\int\psi^2(x,t)\{\alpha_F\cdot\psi(x,t)\}dF(x).$$

Notice that this approximation works as well as (6.3) in the tails; see section 6.2.

We now apply Definition 7.1 to (7.4) and compute the tail area influence function. To achieve this we have to evaluate the tail area under the contaminated distribution $F_{\epsilon,x}=(1-\epsilon)F+\epsilon\Delta_x$. Although this distribution formally does not satisfy the regularity conditions required for (7.4) to hold (see sections 4.2 and 6.2), one can first consider a mixture $(1-\epsilon)F+\epsilon G_\delta$, where $G_\delta$ has a density and the property that as $\delta$ approaches to 0, $G_\delta$ approaches $\Delta_x$. Then, (7.4) holds for the mixture and a limiting argument can be used to show that it holds for $F_{\epsilon,x}$. Using the defining equation (7.5) for $\alpha$ we obtain

$$(\partial/\partial\epsilon)c_{F_{\epsilon,x}}^{-n}(t)\Big|_{\epsilon=0}=nc_F^{-n}(t)\cdot\left[c_F(t)\cdot\exp(\alpha_F(t)\cdot\psi(x,t))-1\right]$$

and

$$\begin{aligned}(\partial/\partial\epsilon)&\left[c_{F_{\epsilon,x}}^{-n}(t)/(\sigma_{F_{\epsilon,x}}(t)\cdot\alpha_{F_{\epsilon,x}}(t))\right]_{\epsilon=0}\\ &=\{c_F^{-n}(t)/\left[\sigma_F(t)\cdot\alpha_F(t)\right]\}\cdot\{n\left[c_F(t)\cdot\exp(\alpha_F(t)\cdot\psi(x,t))-1\right]\\ &\quad-(\dot\alpha_F/\alpha_F)(t)-(\dot\sigma_F/\sigma_F)(t)\},\end{aligned}$$

where the dot denotes the derivative with respect to $\epsilon$ at $\epsilon = 0$. Therefore, the tail area influence function of a M-statistic defined by a function $\psi$ (see (7.3)) is given by

$$TAIF(x;t;\psi,F) = (2\pi)^{-1/2}\{c_F^{-n}(t)/[\sigma_F(t)\cdot\alpha_F(t)]\}\cdot$$
$$\{n^{1/2}[c_F(t)\cdot\exp(\alpha_F(t)\cdot\psi(x,t)) - 1] + 0(n^{-1/2})\}. \qquad (7.6)$$

The $TAIF$ can be interpreted in the same way as any influence function; see Hampel et al. (1986), chapters 2, 3. For instance, if $K_n^{(a)}$ is the critical value of a test of nominal level $a$ based on (7.3), $TAIF\ (x;K_n^{(a)};\psi,F_{\theta_0})$ is the influence function of the level of this test and describes the robustness of validity of the test. A bounded (in $x$) $TAIF\ (x;K_n^{(a)};\psi,F_{\theta_0})$ indicates that the maximum influence of a single observation on the level of the test is bounded. Hence, the test has robustness of validity. Similarly for the power and robustness of efficiency.

In other words, $TAIF$ when applied to a testing problem, can be viewed as a small sample refinement of the (asymptotic) level influence function ($LIF$) and the (asymptotic) power influence function ($PIF$) discussed in Hampel et al. (1986), chapter 3. In fact the following theorem holds.

**Theorem 7.1**

Given $0 < a < 1$ (level), let $K_n^{(a)}$ (critical value) be defined by $P_{F_{\theta_0}}(T_n > K_n^{(a)}) = a$. Define a sequence $\theta_n$ of alternatives as $\theta_n = \theta_0 + \delta n^{-1/2}$, $\delta > 0$.

Then, under the assumptions A4.1-A4.5 of section 4.2 the following holds:

*(i)*
$$\lim_{n\to\infty} n^{-1/2}TAIF(x;K_n^{(a)};\psi,F_{\theta_0})$$
$$= \phi(\Phi^{-1}(1-a))\cdot\psi(x,\theta_0)/(E\psi^2)^{1/2} = \text{LIF}(x;\psi,F_{\theta_0}),$$

*(ii)*
$$\lim_{n\to\infty} n^{-1/2}TAIF(x;K_n^{(a)};\psi,F_{\theta_n})$$
$$= \phi(\Phi^{-1}(1-a) - \delta|E\psi'|/(E\psi^2)^{1/2})\cdot\psi(x;\theta_0)/(E\psi^2)^{1/2}$$
$$= PIF(x;\psi,F_{\theta_0}),$$

where $\phi$ and $\Phi$ are the density and the cumulative of the standard normal distribution, respectively, and $E\psi' = \int\psi'(x,\theta_0)dF_{\theta_0}(x)$, $E\psi^2 = \int\psi^2(x,\theta_0)dF_{\theta_0}(x)$ and $\psi'$ denotes differentiation with respect to $\theta$.

Proof: We prove (i). More tedious computations with similar arguments lead to (ii). By (7.6) we have to compute

$$\lim_{n\to\infty}(2\pi)^{-1/2}\left[c_{F_{\theta_0}}^{-n}(\cdot)/(\sigma_{F_{\theta_0}}(\cdot)\alpha_{F_{\theta_0}}(\cdot))\right]\cdot$$
$$\left[c_{F_{\theta_0}}(\cdot)\exp(\alpha_{F_{\theta_0}}(\cdot)\cdot\psi(x,\cdot)) - 1 + 0(n^{-1})\right], \qquad (7.7)$$

where the argument $(\cdot)$ of the functions involved is $K_n^{(a)}$. From the asymptotic normality of $T_n$ (see Huber 1967, 1981)

$$L_{F_{\theta_0}}(n^{1/2}(T_n - \theta_0)) \to N(0, E\psi^2/(-E\psi')^2),$$

as $n \to \infty$, we obtain

$$K_n^{(a)} = \theta_0 + n^{-1/2}\Phi^{-1}(1-a)\cdot\left[E\psi^2/(E\psi')^2\right]^{1/2} + 0(n^{-1}). \tag{7.8}$$

Using the defining equations for $c$, $\alpha$, $\sigma$ (see (7.5)) and the Fisher-consistency of $T$, that is $T(F_{\theta_0}) = \theta_0$, we get

$$K_n^{(a)} \to \theta_0,\ \ \alpha_{F_{\theta_0}}(K_n^{(a)}) \to 0,\ \ c_{F_{\theta_0}}(K_n^{(a)}) \to 1,\ \ \sigma^2_{F_{\theta_0}}(K_n^{(a)}) \to E\psi^2,$$

when $n \to \infty$.

Now define $B_n := \log c^{-n}_{F_{\theta_0}}(K_n^{(a)})$.

$$\lim_{n\to\infty} B_n = -\lim_{n\to\infty}\left[\log c_{F_{\theta_0}}(K_n^{(a)})\right]/n^{-1}$$

and by L'Hôpital's rule and (7.8)

$$= -\frac{1}{2}\Phi^{-1}(1-a)\cdot\left[E\psi^2/(E\psi')^2\right]^{1/2}\cdot\lim_{n\to\infty}(c'_{F_{\theta_0}}/c_{F_{\theta_0}})(K_n^{(a)})/n^{-1/2}, \tag{7.9}$$

where $c'_{F_{\theta_0}}(t) = (\partial/\partial t)c_{F_{\theta_0}}(t)$. By the defining equation (7.5) we have

$$c'_{F_{\theta_0}}(t) = -c_{F_{\theta_0}}(t)\cdot\alpha_{F_{\theta_0}}(t)\cdot A_{F_{\theta_0}}(t),$$

where $A_F(t) = c_F(t)\int\psi'(x,t)\exp\{\alpha_F\cdot\psi(x,t)\}dF(x)$, and $A_{F_{\theta_0}}(K_n^{(a)}) \to E\psi'$, as $n \to \infty$. Therefore, from (7.8) we obtain

$$\lim_{n\to\infty} B_n = \frac{1}{2}\Phi^{-1}(1-a)\cdot(E\psi^2)^{1/2}sgn(E\psi')\cdot\lim_{n\to\infty}\alpha_{F_{\theta_0}}(K_n^{(a)})/n^{-1/2}. \tag{7.10}$$

Since (dropping the arguments)

$$\alpha' = -\left[\int\psi'\exp(\alpha\cdot\psi)dF + \alpha\cdot\int\psi\cdot\psi'\exp(\alpha\cdot\psi)dF\right]/\int\psi^2\cdot\exp(\alpha\cdot\psi)dF,$$

using L'Hôpital's rule in (7.10), we have

$$\lim_{n\to\infty} B_n = -\frac{1}{2}\left[\Phi^{-1}(1-a)\right]^2, \tag{7.11}$$

and finally

$$\lim_{n\to\infty} c^{-n}_{F_{\theta_0}}(K_n^{(a)}) = (2\pi)^{1/2}\phi(\Phi^{-1}(1-a)).$$

Therefore,

$$\lim_{n\to\infty} n^{-1/2}TAIF(x;K_n^{(a)};\psi,F_{\theta_0})$$

$$= \phi(\Phi^{-1}(1-a))/(E\psi^2)^{1/2}\cdot\lim_{n\to\infty}\left[c_{F_{\theta_0}}(\cdot)\cdot\exp(\alpha_{F_{\theta_0}}(\cdot)\cdot\psi(x,\cdot)) - 1\right]/\alpha_{F_{\theta_0}}(\cdot)$$

$$= \left[\phi(\Phi^{-1}(1-a))\right]/(E\psi^2)^{1/2}\cdot\psi(x,\theta_0),$$

and this equals $LIF(x;\psi,F_{\theta_0})$ as defined in Hampel et al. (1986), (3.2.13), p. 199 and p. 204.

□

### 7.3.c Approximation of The Maximum Level and Minimum Power Over Gross-error Models

Influence functions can be used to extrapolate the value of the functional of interest over a neighborhood of a given model distribution F; cf. Hampel et al. (1986), p. 173, 175, 200. Consider, for instance, the gross-error model

$$P_\epsilon(F) := \{G : G = (1-\epsilon)F + \epsilon H, \ \ H \text{ arbitrary distribution}\}.$$

Using the first two terms of the von Mises expansion (cf. von Mises 1947, Hampel et al. 1986, p. 85) we obtain

$$P_{(1-\epsilon)F+\epsilon H}(T_n > t) \cong P_F(T_n > t) + \epsilon \int TAIF(x;t;T_n,F)dH(x),$$

hence

$$s(t;T_n,F,\epsilon) := sup_H P_{(1-\epsilon)F+\epsilon H}(T_n > t) \cong P_F(T_n > t) + \epsilon \cdot sup_x TAIF(x;t;T_n,F). \tag{7.12}$$

Similarly for the infimum we get

$$i(t;T_n,F,\epsilon) := inf_H P_{(1-\epsilon)F+\epsilon H}(T_n > t) \cong P_F(T_n > t) + \epsilon \cdot inf_x TAIF(x;t;T_n,F). \tag{7.13}$$

(7.12) and (7.13) require the approximation to be valid uniformly over shrinking $\epsilon$-contamination neighbourhoods. For M-statistics defined by (7.3), this can be obtained by imposing some additional conditions on the function $\psi$ as in Rieder (1980), p. 114. In general these approximations will be valid only when the amount of contamination is much smaller than the breakdown point of $T_n$.

This type of approximation has been used successfully for other kinds of problems, for instance the approximation of the variance of an estimator over $\epsilon$-contaminated models, and leads to very accurate results (cf. Hampel 1983; Hampel et al. 1986, p. 173).

Let us now apply (7.12) and (7.13) to the testing problem. Set $t = K_n^{(a)}$,the critical value of a level-a test, $F = F_{\theta_0}$ in (7.12) and $F = F_\theta$, $\theta \neq \theta_0$, in (7.13). Consider the class of tests based on a M-statistic $T_n$ defined by a function $\psi$ (see (7.3)) and approximate $TAIF$ using (7.6). To compare with asymptotic results we replace $\epsilon$ by $\epsilon_n = \epsilon \cdot n^{-1/2}$ and we define a sequence of alternatives $\theta_n = \theta_0 + \delta n^{-1/2}$. Then we have

$$\begin{aligned}\text{supremum level } &= s(K_n^{(a)};\psi,F_{\theta_0},\epsilon_n)\\ &\cong a + \epsilon \cdot b_n(K_n^{(a)},F_{\theta_0}) \cdot \{c_{F_{\theta_0}}(K_n^{(a)}) \cdot \\ &\qquad \exp\left[\alpha_{F_{\theta_0}}(K_n^{(a)}) \cdot sup_x \psi(x,K_n^{(a)})\right] - 1\}\end{aligned} \tag{7.14}$$

and

$$\begin{aligned}\text{infimum power } &= i(K_n^{(a)};\psi,F_{\theta_n},\epsilon_n)\\ &\cong \beta(\theta_n) + \epsilon \cdot b_n(K_n^{(a)},F_{\theta_n}) \cdot \{c_{F_{\theta_n}}(K_n^{(a)}) \cdot \\ &\qquad \exp\left[\alpha_{F_{\theta_n}}(K_n^{(a)}) \cdot inf_x \psi(x,K_n^{(a)})\right] - 1\},\end{aligned} \tag{7.15}$$

where $b_n(t,F) := (2\pi)^{-1/2} \cdot c_F^{-n}(t)/[\sigma_F(t) \cdot \alpha_F(t)]$ and $\beta(\theta_n)$ is the power of the test at the alternative $F_{\theta_n}$. Since, in general the exact value of $K_n^{(a)}$ is difficult to compute, we can use the approximation given by (7.8).

According to Theorem 7.1, as $n \to \infty$, (7.14) becomes

$$\lim_{n\to\infty} s(K_n^{(a)}; \psi, F_{\theta_0}, \epsilon_n) \cong a + \epsilon \cdot sup_x LIF(x; \psi, F_{\theta_0}), \tag{7.16}$$

and (7.15),

$$\lim_{n\to\infty} i(K_n^{(a)}; \psi, F_{\theta_0}, \epsilon_n) \cong \beta_{as} + \epsilon \cdot inf_x PIF(x; \psi, F_{\theta_0}), \tag{7.17}$$

where $\beta_{as} = 1 - \Phi(\Phi^{-1}(1-a) - \delta[(E\psi')^2/E\psi^2]^{1/2})$ is the asymptotic power of the test at the model.

As an illustration we compare the exact results with the approximations (7.14) and (7.15) (and the approximations obtained from $LIF$ and $PIF$ based on the asymptotic distribution) in the following situation:

$$F_\theta(x) = \Phi(x - \theta),$$
$$H_0 : \theta = 0,$$
$$T_n = \text{ median, that is } \psi(x,t) = sgn(x-t).$$

We choose this case because it allows an explicit exact computation of the supremum of the level and the infimum of the power. Details of computations can be found in Field and Ronchetti (1985).

| $n$ | $K_n^{(a)}$ | $\epsilon$ | 0.00 | .05 | .01 | .10 |
|---|---|---|---|---|---|---|
| 3 | 1.101 | | 5.00 | 5.36 | 6.89 | 9.03 |
| 5 | .880 | | | 5.26 | 6.37 | 7.94 |
| 7 | .754 | | | 5.22 | 6.17 | 7.49 |
| 11 | .609 | | | 5.19 | 5.98 | 7.08 |

**Exhibit 7.4**

Exact supremum of the level $s(K_n^{(a)}; med, \Phi, \epsilon_n)$ (in %) over $(1-\epsilon_n) \cdot \Phi(x) + \epsilon_n H(x)$ of the test based on the median ($a = 5\%$)

| $n$ | $K_n^{(a)}$ (approx.) | $\epsilon$ | 0.00 | .01 | .05 | .10 |
|---|---|---|---|---|---|---|
| 3 | 1.190 | | 5.00 | 5.34 | 6.72 | 8.43 |
| 5 | .922 | | | 5.25 | 6.24 | 7.48 |
| 7 | .779 | | | 5.21 | 6.06 | 7.11 |
| 11 | .622 | | | 5.18 | 5.90 | 6.79 |
| 100 | .206 | | | 5.12 | 5.61 | 6.22 |
| 10000 | .021 | | | 5.10 | 5.52 | 6.05 |
| $\infty$ | 0.000 | | | 5.10 | 5.51 | 6.03 |

**Exhibit 7.5**

Approximate supremum of the level $s(K_n^{(a)}; med, \Phi, \epsilon_n)$ (in %) over $(1-\epsilon_n)\Phi(x) + \epsilon_n H(x)$ of the test based on the median ($a = 5\%$) (approx. (7.14), (7.16)).

| $n$ | $K_n^{(a)}$ | $\delta$ $\epsilon$ | 0.00 | .01 | .05 | .10 |
|---|---|---|---|---|---|---|
| 3 | 1.101 | .5 | 11.20 | 11.08 | 10.61 | 10.04 |
| | | 1.0 | 21.62 | 21.41 | 20.54 | 19.48 |
| | | 1.5 | 36.21 | 35.87 | 34.52 | 32.85 |
| 5 | .880 | .5 | 10.95 | 10.83 | 10.34 | 9.75 |
| | | 1.0 | 20.85 | 20.63 | 19.77 | 18.71 |
| | | 1.5 | 34.72 | 34.38 | 33.07 | 31.45 |
| 7 | .754 | .5 | 10.85 | 10.72 | 10.22 | 9.60 |
| | | 1.0 | 20.53 | 10.31 | 19.43 | 18.36 |
| | | 1.5 | 34.09 | 33.76 | 32.46 | 30.85 |
| 11 | .609 | .5 | 10.78 | 10.64 | 10.11 | 9.47 |
| | | 1.0 | 20.28 | 20.05 | 19.15 | 18.06 |
| | | 1.5 | 33.58 | 33.26 | 31.94 | 30.34 |

**Exhibit 7.6**

Exact infimum of the power $i(K_n^{(a)}; med, \phi, (\cdot - \theta_n), \epsilon_n)$ (in %) over $(1-\epsilon_n)\cdot\Phi(x-\theta_n)+\epsilon_n H(x)$ of the test based on the median ($a = 5\%$).

| $n$ | $K_n^{(a)}$ (approx.) | $\delta$ | $\epsilon$ 0.00 | .01 | .05 | .10 |
|---|---|---|---|---|---|---|
| 3 | 1.190 | .5 | 10.15 | 10.05 | 9.67 | 9.19 |
| | | 1.0 | 19.26 | 19.08 | 18.38 | 17.49 |
| | | 1.5 | 32.26 | 31.98 | 30.85 | 29.44 |
| 5 | .922 | .5 | 10.34 | 10.23 | 9.79 | 9.24 |
| | | 1.0 | 19.49 | 19.29 | 18.52 | 17.54 |
| | | 1.5 | 32.43 | 32.13 | 30.95 | 29.46 |
| 7 | .779 | .5 | 10.42 | 10.30 | 9.83 | 9.24 |
| | | 1.0 | 19.59 | 19.38 | 18.56 | 17.53 |
| | | 1.5 | 32.51 | 32.20 | 30.98 | 29.45 |
| 11 | .622 | .5 | 10.50 | 10.37 | 9.85 | 9.21 |
| | | 1.0 | 19.68 | 19.46 | 18.59 | 17.50 |
| | | 1.5 | 32.57 | 32.26 | 30.99 | 29.41 |
| 100 | .206 | .5 | 10.62 | 10.46 | 9.81 | 8.99 |
| | | 1.0 | 19.82 | 19.57 | 18.54 | 17.26 |
| | | 1.5 | 32.68 | 32.33 | 30.95 | 29.23 |
| 10000 | .021 | .5 | 10.63 | 10.45 | 9.73 | 8.82 |
| | | 1.0 | 19.84 | 19.56 | 18.46 | 17.08 |
| | | 1.5 | 32.69 | 32.33 | 30.90 | 29.10 |
| $\infty$ | .000 | .5 | 10.64 | 10.45 | 9.72 | 8.80 |
| | | 1.0 | 19.85 | 19.57 | 18.45 | 17.06 |
| | | 1.5 | 32.70 | 32.34 | 30.90 | 29.09 |

**Exhibit 7.7**

Approximate infimum of the power $i(K_n^{(a)}; med, \Phi(\cdot - \theta_n), \epsilon_n)$ (in %) over $(1-\epsilon_n)\Phi(x-\theta_n) + \epsilon_n H(x)$ of the test based on the median ($a = 5\%$) (approx. (7.13), (7.15)).

A comparison of Exhibit 7.4 and Exhibit 7.5 shows the general good accuracy of the approximation based on $TAIF$ down to small $n$ $(= 3, 5)$. Moreover, this approximation improves the asymptotic result based on $LIF$. For instance, for $\epsilon = 10\%$ and $n = 3$, the exact supremum of the level is 9.03%, the approximation based on $TAIF$ gives 8.43%, while the asymptotic normality predicts 6.03%.

At the moment the advantage of the asymptotic result lies in its greater flexibility: the expression of the "asymptotic" influence function is simpler and can be computed for general classes of tests. Therefore, optimal robust tests can be derived using the asymptotic theory and eventually refined for small sample sizes.

The results related to the infimum of the power (Exhibits 7.6 and 7.7) seem to indicate that here the accuracy of the approximation based on $TAIF$ is not as good as for the supremum of the level. The reason seems to lie in the approximation of the power $\beta(\theta)$

at the model. With a better correction term it is possible to achieve the high accuracy obtained for the supremum of the level.

For example, if we used in (7.15) the exact value for $\beta(\theta_n)$ instead of the approximation $\beta_{as}$, we would obtain for $n = 3$, $\delta = .5$ and $\epsilon = .01, .05, .10$ the following values for the given power: 11.10%, 10.72%, 10.24%.

### 7.3.d Robustness Versus Efficiency

Given a test based on a test statistic $T_n$ defined by $\psi$, one can compute up to terms of the order $n^{-1/2}$ the tail area influence function (7.6) and discuss the behaviour of $TAIF$ with respect to $x$, $t$, $n$. One will strive for a $TAIF$ that is bounded (not necessarily in a symmetric way) with respect to $x$ to limit the influence of outliers, that is continuous in $x$ to limit grouping and rounding effects, etc. Moreover, putting $F = F_\theta$ and $t = K_n^{(a)}$, the critical value of a test, by means of $TAIF$ one can investigate the influence of outliers on the level and on the power and their behaviour over gross-error models (see subsection 7.3b).

A natural problem that arises is to try and find a balance between robustness and efficiency. We can look for a test in the class (7.3) that maximizes the power, under the condition of a bounded (with respect to $x$) $TAIF$ $(x; K_n^{(a)}; \psi, F_\theta)$. Using (7.4) and (7.6) to approximate the power and the tail area influence function, one can hope to find a function $\psi_{opt}(x,t)$ that satisfies a first order condition for the optimality problem. We conjecture that in the normal-location case, that is, $F_\theta(x) = \Phi(x-\theta)$, $\psi_{opt}(x,t)$ equals the Huber-function

$$\begin{aligned} \psi_k(x-t) &= x-t && \text{if } |x-t| \le k \\ &= k \ sgn(x-t) && \text{otherwise,} \end{aligned}$$

although an exact proof seems to be rather complicated.

## 7.4. SADDLEPOINT APPROXIMATION FOR THE WILCOXON TEST

In this section we discuss an application of the method of steepest descent to rank procedures. More precisely, we discuss the approximation of the density and tail areas of the Wilcoxon test statistic.

Given a sample of $n$ independent observations $x_1, \cdots, x_n$, we want to test the null hypothesis $H_0$ that these observations have a common symmetric distribution about 0. To this end one can use the Wilcoxon test or signed-rank test which is defined by the test statistic

$$T_n = \sum_{i=1}^{n} R_i \cdot sgn(x_i), \tag{7.18}$$

where $R_i$ is the rank of $x_i$ according to its absolute value. If large values of $T_n$ are significant, we are then interested in the tail probability

$$P = P[T_n \ge t_0 | H_0]. \tag{7.19}$$

Although tables of the exact cumulative distribution

$$q_k = P[0 \le T_n \le k | H_0] \tag{7.20}$$

are available, it is nevertheless interesting to investigate the performance of saddlepoint approximations in the case. Moreover, a good approximation of (7.19) has its own interest because it would allow to compute P-values for *any* given $n$ and $t_0$. This is even more important in the two-sample case, where two parameters (the sample sizes of the two groups) are involved in addition to $t_0$.

There are different ways one can use to approximate (7.19). For instance, one could apply the general saddlepoint approximation developed in section 4.3. By using an Edgeworth expansion to approximate the cumulant generating function of $T_n$, one can approximate by means of the techniques of section 4.3 a saddlepoint expansion to the density or a saddlepoint expansion to the tail area. However, it turns out that in this case the exact probability generating function is known and therefore a direct saddlepoint approximation can be computed. This was done by Helstrom (1986b) and we follow here his development. Similar results can be found in Robinson et al. (1988).

Since $T_n$ can assume only discrete values, define $p_j = P\big[T_n = j|H_0\big]$. Then, by Cauchy theorem

$$p_j = (2\pi)^{-1} \int_{\mathcal{P}} z^{-(j+1)} h_n(z) dz,$$

$$q_k = P\big[0 \le T_n \le k|H_0\big] = \sum_{j=0}^{k} p_j = (2\pi i)^{-1} \int_{\mathcal{P}} \Big\{ \sum_{j=0}^{k} z^{-(j+1)} \Big\} h_n(z) dz$$

$$= (2\pi i)^{-1} \int_{\mathcal{P}} z^{-(k+1)} (1-z)^{-1} h_n(z) dz, \tag{7.21}$$

where $h_n(z) = \sum_{j=0}^{\infty} p_j z^j = \prod_{\ell=1}^{n} (1+z^{\ell})/2$, and $\mathcal{P}$ is any closed path surrounding the origin in the complex plane, but not the point $z = 1$. By defining

$$\tilde{w}_n(z) = \log h_n(z) - (k+1)\log z - \log(1-z),$$

the right hand side of (7.21) assumes the form of the integral (3.1), where the "large parameter" $v(=n)$ is already included in $\tilde{w}_n(z)$. At this point Helstrom (1986b) computes numerically the saddlepoint $z_0$ which satisfies

$$\tilde{w}_n'(z_0) = 0,$$

and the path of steepest descent given by

$$\mathcal{I}\tilde{w}_n(z) \equiv 0,$$

and he integrates along this path by using the trapezoidal rule; cf. Rice (1973). We refer to that paper for the numerical aspects.

Exhibit 7.8 gives the relative error $(= (q_k(\text{SAD}) - q_k(\text{exact}))/q_k(\text{exact}))$ for several sample sizes and probability levels. These results show once more the great accuracy of the saddlepoint approximation even far out in the tail. Similar results are obtained for the two-sample case; see Helstrom (1986b).

**Probability Level**

| | 0.00005 | | 0.0005 | | 0.005 | |
|---|---|---|---|---|---|---|
| **Sample Size** | $N_s$ | **Rel. Error** | $N_s$ | **Rel. Error** | $N_s$ | **Rel. Error** |
| 20 | 18 | -0.27(-2) | 16 | -0.82(-3) | 14 | 0.15(-4) |
| | 35 | -0.27(-2) | 31 | -0.81(-3) | 27 | 0.15(-4) |
| 25 | 20 | -0.17(-3) | 18 | -0.71(-5) | 16 | 0.58(-5) |
| | 40 | -0.13(-3) | 36 | -0.74(-5) | 32 | 0.57(-5) |
| 30 | 23 | 0.44(-5) | 21 | -0.14(-5) | 17 | 0.13(-6) |
| | 45 | 0.37(-5) | 41 | -0.14(-5) | 31 | -0.19(-7) |
| 35 | 25 | -0.55(-6) | 18 | -0.93(-7) | 17 | 0.16(-6) |
| | 50 | -0.63(-6) | 36 | -0.98(-7) | 33 | -0.14(-7) |
| 40 | 21 | -0.44(-7) | 18 | 0.22(-8) | 17 | 0.20(-6) |
| | 42 | -0.45(-7) | 35 | -0.74(-8) | 33 | -0.17(-8) |

**Probability Level**

| | 0.01 | | 0.05 | | 0.2 | |
|---|---|---|---|---|---|---|
| **Sample Size** | $N_s$ | **Rel. Error** | $N_s$ | **Rel. Error** | $N_s$ | **Rel. Error** |
| 20 | 13 | 0.22(-4) | 12 | -0.30(-5) | 11 | 0.56(-4) |
| | 26 | 0.21(-4) | 24 | -1.00(-5) | 22 | -0.44(-6) |
| 25 | 15 | -0.83(-6) | 14 | 0.44(-5) | 9 | 0.56(-4) |
| | 30 | -0.12(-5) | 28 | 0.57(-6) | 18 | -0.29(-6) |
| 30 | 16 | 0.23(-6) | 16 | 0.51(-5) | 8 | 0.55(-4) |
| | 29 | -0.20(-6) | 32 | -0.18(-7) | 15 | -0.11(-8) |
| 35 | 17 | 0.48(-6) | 9 | 0.53(-5) | 7 | 0.55(-4) |
| | 33 | 0.12(-7) | 17 | -0.34(-8) | 14 | 0.54(-9) |
| 40 | 11 | 0.55(-6) | 8 | 0.56(-5) | 7 | 0.55(-4) |
| | 19 | -0.11(-8) | 15 | -0.33(-9) | 14 | 0.47(-9) |

**Exhibit 7.8** (from Helstrom, 1986b)

Relative error $(= (q_k(\text{SAD}) - q_k(\text{exact}))/q_k(\text{exact})$ for the distribution of the one-sample Wilcoxon statistic. $N_s$ is the number of steps in the numerical integration involved. 0.22(−4) means $0.22 \cdot 10^{-4}$.

## 7.5. APPLICATIONS IN COMMUNICATION THEORY AND RADAR DETECTION

In signal detection problems engineers work with very small false-alarm probabilities. Accurate real time approximations of extreme tail areas are therefore required in these situations; cf. Exhibit 2.3. There is a growing number of papers in the engineering literature where the method of steepest descent and saddlepoint techniques are used to find good approximations to tail areas. In this section we summarize the key ideas with two typical examples.

The first example is taken from Helstrom and Ritcey (1984).

Consider a receiver that integrates a large number $n$ of pulses. The problem of radar detection of a known signal in additive white Gaussian noise can be formulated as a testing problem

$$H_0 : T_n = \frac{1}{2}\sum_{j=1}^{n}(x_j^2 + y_j^2),$$

against

$$H_1 : T_n = \frac{1}{2}\sum_{j=1}^{n}[(x_j + s_j)^2 + (y_j + t_j)^2], \tag{7.22}$$

where $x_1, \cdots, x_n, y_1, \cdots, y_n$ are $n$ iid Gaussian random variables with zero mean and unit variance and $s_j^2 + t_j^2 = |d_j|^2$. The $d_j = s_j + it_j$ are the complex amplitudes of the target echoes and the total signal-to-noise ratio is $S = \frac{1}{2}\sum_{j=1}^{n}|d_j|^2$. The $d_j$'s are fixed for a steady target and are random variables when the target fluctuates. In the latter case the phase of the amplitude is uniformly distributed on $[0, 2\pi]$. According to the Neyman-Pearson Lemma the system decides $H_0$ or $H_1$ according to whether

$$T_n \underset{H_0}{\overset{H_1}{\gtrless}} t_0,$$

where $t_0$ is the critical value depending on the false-alarm probability (level of the test). Note that the test based on this test statistic is not robust and $T_n$ can be modified to be resistant to outlying observations. However, our goal here is to investigate the distribution of $T_n$.

No simple analytic expression is available for the density of the test statistic $T_n$ especially in the case of a fluctuating target. Moreover, the computation of $t_0$ for very small false-alarm probabilities and the computation of the detection probability (power of the test) requires the integration of the density of $T_n$ far out in the tails. In this case the normal approximation is of little help; cf. Exhibit 2.3.

In many electrical engineering applications, although the density in general is difficult to evaluate, the moment generating function is easy to determine. From Helstrom and Ritcey (1984) the moment generating function for a nonfluctuating target (for a fixed signal-to-noise ratio $S$) is

$$M_n(\alpha|S) = (1-\alpha)^{-n}\exp\{S\alpha/(1-\alpha)\}$$

and that for a fluctuating target is

$$M_n(\alpha) = (1-\alpha)^{-n} M_n^{(S)}(-\alpha/(1-\alpha)),$$

where $M_n^{(S)}$ is the (known) moment generating function of $S$. Then, by the same arguments as in section 3.3 or section 4.3, one can write the density of $T_n$ as

$$f_n(t) = \frac{1}{2\pi i} \int_{\tau - i\infty}^{\tau + i\infty} M_n(z) \exp(-zt) dz \tag{7.23}$$

and the false-alarm probability

$$\int_{t_0}^{\infty} f_n(t) dt = \frac{1}{2\pi i} \int_{\tau - i\infty}^{\tau + i\infty} M_n(z) \exp(-zt_0) dz \tag{7.24}$$

cf. also (6.1).
The detection probability can be computed in a similar way.

At this point the integrals in (7.23) and (7.24) could be evaluated by using the saddlepoint techniques discussed in the previous chapters. However, engineers rewrite the integral in (7.24) as $\int_{\tau - i\infty}^{\tau + i\infty} \exp\left[\tilde{w}_n(z)\right] dz$, where $\tilde{w}_n(z) = -\log z - zt_0 + K_n(z)$, with $K_n(z) = \log M$, and determine the saddlepoint $z_0$ and the corresponding path of steepest descent. Now, instead of developing the exponent in a series around the saddlepoint and keeping the leading term, they evalute numerically the integral along the contour of integration defined in the x-y plane by $\Im w_n(x+iy) = 0$. This contour corresponds to the path of steepest descent from the saddlepoint $z_0$; see section 3.2. The numerical integration is performed using the trapezoidal rule as developed in Rice (1973). Since the contour of integration is a curve on the x-y plane defined implicitly, it is often approximated by a parabola crossing the real axis at the (real) saddlepoint $z_0$.

Exhibit 7.9 shows a comparison of saddlepoint approximations and numerical integration techniques for some selected cases. The integration's contours considered in this exhibit are the "vertical contour" and the "parabolic contour". The former is obtained by approximating the exact integration's contour by a vertical straight line at the saddlepoint whereas the latter is obtained by using a quadratic approximation of the integration path at the saddlepoint. Exhibit 7.9 shows the improvement of the accuracy by using the numerical integration and the faster convergence of the parabolic contour method compared to the vertical one.

| $\bar{S}/n$ in dB | Saddlepoint approximation | Numerical contour integration: Vertical contour | # steps | Parabolic contour | # steps |
|---|---|---|---|---|---|
| | | $n = 10$, $\kappa = 20$, $t_0 = 32.717$ | | | |
| $-\infty$ | 1.0008(−6)* | 9.971482(−7)* | 16 | 9.951845(−7)* | 5 |
| | | 9.951150(−7)* | 32 | 9.951149(−7)* | 10 |
| 5.0 | 2.1588(-1) | 2.190677(-1) | 12 | 2.186425(-1) | 6 |
| | | 2.184155(-1) | 24 | 2.184151(-1) | 11 |
| | | $n = 100$, $\kappa = 200$, $t_0 = 154.9$ | | | |
| $-\infty$ | 1.0064(−6)* | 1.007402(−6)* | 6 | 1.007217(−6)* | 5 |
| | | 1.007099(−6)* | 11 | 1.007099(−6)* | 10 |
| -2.3 | 3.8960(-1) | 4.100784(-1) | 7 | 4.097784(-1) | 6 |
| | | 4.086649(-1) | 13 | 4.086641(-1) | 12 |
| 0.0 | 4.9742(-3) | 4.993271(-3) | 6 | 4.991731(-3) | 5 |
| | | 4.990710(-3) | 11 | 4.990710(-3) | 10 |
| | | $n = 500$, $\kappa = 1000$, $t_0 = 613.576$ | | | |
| $-\infty$ | 9.9900(−7)* | 1.000187(−6)* | 5 | 1.000160(−6)* | 5 |
| | | 1.000041(−6)* | 10 | 1.000041(−6)* | 10 |
| -7.0 | 2.8300(−1)* | 2.989386(−1)* | 6 | 2.990300(−1)* | 6 |
| | | 2.982982(−1)* | 12 | 2.982983(−1)* | 12 |
| -5.0 | 5.8950(-2) | 5.985495(-2) | 6 | 5.984295(-2) | 6 |
| | | 5.980515(-2) | 11 | 5.980515(-2) | 11 |

**Exhibit 7.9**
(from Helstrom and Ritcey (1984))

False dismissal probability (type II error probabilty) for radars with fluctuating target (Swerling Case IV Target) computed by means of saddlepoint techniques and numerical integration on two approximation of the contour (vertical contour and parabolic contour).

$\bar{S}$ is the average signal-to-noise ratio and $\bar{S}/n$ is given in $dB$ ($= 10\log_{10}(\cdot)$). $\kappa = 2n$ is the parameter of the distribution of $S$ for Swerling Case IV Targets. 5.8950(−2) means $5.8950 \cdot 10^{-2}$. An asterisk indicates that 1 - false dismissal probability (= detection probability = power of the test) is reported.

The second example is taken from Helstrom (1986a).

Consider a binary symmetric channel with intersymbol interference. The output of the receiver at time $t$ has the form

$$x(t) = s(t) + \epsilon(t), \tag{7.25}$$

where the signal $s(t)$ is given by

$$s(t) = \sum_{j=-\infty}^{+\infty} b_j a(t - jT)$$

and $\epsilon(t)$ is the noise with mean zero and a symmetric density such as the normal. The $b_j$'s are $+1$ or $-1$ with equal probability and express the transmitted message. They are assumed uncorrelated. $a(t)$ is the pulse transmitted and $T$ is the interval between pulses. By sampling at a time $t$ we obtain

$$x = \sum_{j=-\infty}^{+\infty} b_j a_j + \epsilon, \qquad a_j = a(t - jT).$$

The receiver then decides whether a particular one of the $b_j$, say $b_0$, is $+1$ or $-1$ by choosing $+1$ if $x > 0$ and $-1$ if $x < 0$. The error probability is given by

$$P_{err} = P\big[x = \eta + a_0 + \epsilon < 0 | b_0 = +1\big],$$

where

$$\eta = \sum_{\substack{j=-\infty \\ j\neq 0}}^{+\infty} b_j a_j .$$

The exact calculation of $P_{err}$ is very difficult because all possible combinations of the $b_j = \pm 1$ must be taken into account, $1 \leq |j| \leq \infty$. Even with only a finite number $n$ of pulses before and after $a_0$, the computation of the error probability involves $2^{2n}$ combinations, where $n$ is large in practical applications.

As in the previous example the moment generating function of $x$ can be determined easily. When a finite number $2n$ of pulses is considered to interfere with $b_0$, the moment generating function is given by

$$M_n(\alpha) = \gamma_n(\alpha) M_n^{(\epsilon)}(\alpha) \exp(a_0 \alpha), \tag{7.26}$$

where $M_n^{(\epsilon)}$ is the (known) moment generating function of the noise $\epsilon$ and

$$\gamma_n(\alpha) = E\big[e^{\eta \cdot \alpha}\big] = E\Big[\exp(\alpha \sum_{\substack{j=-n \\ j\neq 0}}^{n} b_j a_j)\Big]$$

$$= \prod_{\substack{j=-n \\ j\neq 0}}^{n} \cosh(a_j \alpha). \tag{7.27}$$

Then, by the same arguments as in (7.23) and (7.24), we can rewrite the density of $x$ as

$$f_n(x) = \frac{1}{2\pi i} \int_{\tau - i\infty}^{\tau + i\infty} M_n(z) e^{-zx} dz$$

and the error probability

$$P_{err} = \int_{-\infty}^{0} f_n(x)dx = \frac{1}{2\pi i}\int_{T-i\infty}^{T+i\infty} M_n(z)\left\{\int_{-\infty}^{0} e^{-zx}dx\right\}dz$$

$$= -\frac{1}{2\pi i}\int_{T-i\infty}^{T+i\infty} z^{-1}M_n(z)dz. \tag{7.28}$$

The integral in (7.28) can now be evaluated by using saddlepoint techniques and numerical contour integration; see Helstrom (1986a).

Further engineering applications of saddlepoint echniques can be found in Helstrom (1978, 1979, 1985), Helstrom and Rice (1984), and Ritcey (1985).

## 7.6. OTHER APPLICATIONS OF THE SADDLEPOINT METHOD

After Daniels (1954) pioneering paper, saddlepoint techniques have been applied successfully to several types of problems. Several applications are already given in the previous chapters. In this section we summarize some further applications, the goal being to give an idea of the diversity of situations where these techniques can be used.

Keilson (1963) applied saddlepoint techniques to find an approximation to the density of the sum of $N$ iid random variables, where $N$ is a Poisson process, whereas Blackwell and Hodges (1959) and Petrov (1965) used these ideas to derive approximations to tail probabilities of the sum of iid random variables. A generalization of Keilson's paper can be found in Embrechts et al. (1985). A computation is given in Exhibit 3.12.

Although saddlepoint methods rely on the existence of the moment generating function, Daniels (1960) shows a case where the technique can be carried out when this condition fails.

The approximation to the density of the serial correlation coefficient is considered in Daniels (1956) and Durbin (1980b). Moreover, Daniels (1982) uses saddlepoint approximations in birth processes. These papers show that these techniques can be applied in non iid problems. A related paper is Bolthausen (1986) where Laplace approximations for Markov processes are discussed.

An important application is considered by Robinson (1982). In this paper tail areas for permutation tests are approximated and by inversion approximate confidence intervals are constructed; cf. also the short discussion in Daniels (1955).

The steepest descent method combined with numerical integration is used by Imhof (1961), Lugannani and Rice (1984), and Helstrom and Ritcey (1985) to approximate the density of quadratic forms of normal variables and noncentral F-distributions.

An application of similar ideas in the Bayesian framework is provided by Tierney and Kadane (1986). Basically, Laplacian techniques are used to approximate the integrals in the numerator and the denominator when computing a posterior expectation. Software to perform these computations is available in XLISP-STAT; see Tierney (1990).

Important applications in econometrics can be found in Phillips (1978) and Holly and Phillips (1979). In the latter paper the density of the k-class estimator in the case of the equation with two endogeneous variables is approximated. A related paper is Chesher, Peters, Spady (1989).

Further examples are given in Good (1957, 1961), Keilson (1965), Barndorff-Nielsen and Cox (1979), and Davison and Hinkley (1988), and Johns (1988).

## 7.7. OUTLOOK

In this monograph we have shown how small sample asymptotic techniques can be successfully applied to many different types of problems. At different places we have pointed to possible open problems and research directions. Clearly, the striking characteristic of these approximations is the great accuracy down to very small sample sizes. But, in spite of the accumulated numerical evidence on many different problems and in spite of the theoretical results on the expansions, the answer to the question as to *why small sample asymptotics does work so well for very small sample sizes* is still not completely satisfactory. More research probably together with the development of new analytic tools to measure this accuracy is needed.

Small sample asymptotics has now reached the point where it can be implemented in computer packages. Both the practitioner and the researcher would benefit greatly from this software development.

## REFERENCES

Albers, W., Bickel, P.J., and van Zwet, W.R.(1976), "Asymptotic Expansions for the Power of Distribution Free Tests in One-Sample Problem", *Annals of Statistics* **4**, 108–156. [2.5]

Andrews, D.F., Bickel, P.J., Hampel, F.R., Huber, P.J., Rogers, W.H., and Tukey, J.W. (1971), *Robust Estimates of Location:Survey and Advances*, Princeton University Press, Princeton, NJ. [4.5]

Bahadur, R.R. (1971), *Some Limit Theorems in Statistics*, Soc. Ind. Appl. Math., Philadelphia. [6.2]

Barndorff-Nielsen, O. (1980), "Conditionality Resolutions", *Biometrika* **67**, 293–310. [5.4]

Barndorff-Nielsen, O. (1983), "On a Formula for the Distribution of the Maximum Likelihood Estimator," *Biometrika* **70**, 343–365. [4.3, 5.4, 5.5]

Barndorff-Nielsen, O. (1984), "On Conditionality Resolution and the Likelihood Ratio for Curved Exponential Families", *Scandinavian Journal of Statistics* **11**, 157-170. [5.5]

Barndorff-Nielsen, O. (1986), "Inference on Full or Partial Parameters Based on the Standardized Signed Log Likelihood Ratio", *Biometrika* **73**, 307–322. [5.5]

Barndorff-Nielsen, O., and Cox, D.R. (1979), "Edgeworth and Saddle-point Approximations with Statistical Applications" (with discussion), *Journal of the Royal Statistical Society*, B, **41**, 279–312. [2.6,7.6]

Barndorff-Nielsen, O., and Cox, D.R. (1989), *Asymptotic Techniques For Use in Statistics*, London: Chapman and Hall. [3.2]

Barndorff-Nielsen, O., and Pedersen, B.V. (1979), "The Bivariate Hermite Polynomials Up to Order Six", *Scandinavian Journal of Statistics* **6**, 127–128. [2.6]

Berry, A.C. (1941), "The Accuracy of the Gaussian Approximation to the Sum of Independent Variates", *Transactions of the American Mathematical Society* **49**, 122–136. [2.2]

Bhattacharya, R.N., and Ghosh, J.K. (1978), "On the Validity of the Formal Edgeworth Expansion", *Annals of Statistics* **6**, 434–451. [2.6, 4.2]

Bhattacharya, R.N., and Rao, R.R. (1976), *Normal Approximation and Asymptotic Expansions*, New York: Wiley. [1.1, 2.2]

Bickel, P.J. (1974), "Edgeworth Expansions in Nonparametric Statistics", *Annals of Statistics* **2**, 1–20. [2.2, 2.5]

Bickel, P.J., Götze, F., and van Zwet, W.R. (1986), "The Edgeworth Expansion For

U-statistics of Degree Two", *Annals of Statistics* **14**, 1463–1484. [2.5]

Bjerve, S. (1977), "Error Bounds for Linear Combinations of Order Statistics", *Annals of Statistics* **5**, 237–255. [2.2]

Blackwell, D., and Hodges, J.L., Jr. (1959), "The Probability in the Extreme Tail of a Convolution", *Annals of Mathematical Statistics* **30**, 1113–1120. [7.6]

Blaesild, P., and Jensen, J.L. (1985), "Saddlepoint Formulas for Reproductive Exponential Models", *Scandinavian Journal of Statistics* **12**, 193–202. [3.3]

Bolthausen, E. (1986), "Laplace Approximations for Markov Process Expectations", *Probability Theory and Related Fields* **72**, 305–318. [7.6]

Callaert, H., and Janssen, P. (1978), "The Berry -Esseen Theorem for U-statistics", *Annals of Statistics* **6**, 417–421. [2.2]

Callaert, H., Janssen, P., and Veraverbeke, N. (1980), "An Edgeworth Expansion for U-statistics", *Annals of Statistics* **8**, 299–312. [2.5]

Chaganty, N.R., and Sethuraman, J. (1985), "Large Deviations Local Limit Theorems for Arbitrary Sequences of Random Variables", *Annals of Probability* **13**, 95–114. [4.3]

Chaganty, N.R., and Sethuraman, J. (1986), "Multidimensional Large Deviation Local Limit Theorems," *Journal of Multivariate Analysis* **20**, 190-204. [4.3]

Chan, Y., and Wierman, J. (1977), "On the Berry-Esseen Theorem for U-statistics", *Annals of Probability* **5**, 136–139. [2.2]

Charlier, C.V.L. (1905), "Uber das Fehlergesetz", *Ark. Mat. Astr. och Fys.* **2**, 9 pp. [2.3]

Charlier, C.V.L. (1906), "Uber die Darstellung Willkürlicher Funktionen, *Ark. Mat. Astr. och Fys.* **20**, 35 pp. [2.3]

Chebyshev, P.L. (1890), "Sur deux théorèmes relatifs aux probabilités", *Acta Math.* **14**, 305–315. [2.3]

Chernoff, H., Gastwirth, J.L., and Johns, M.V. (1967), "Asymptotic Distribution of Linear Combinations of Functions of Order Statistics With Applications to Estimation," *Annals of Mathematical Statistics* **38**, 52–72. [4.4]

Chesher, A., Peters, S., and Spady, R. (1989), "Approximations to the Distributions of Heterogeneity Tests in the Censored Normal Linear Regression Model," University of Bristol, manuscript. [7.6]

Cicchitelli, G. (1976), "The Sampling Distribution of Linear Combinations of Order Statistics," *Metron* **34**, 269–302. [4.4]

Copson, E.T. (1965), *Asymptotic Expansions*, Cambridge: Cambridge University Press.

[3.2]

Cox, D.R. (1958), "The Regression Analysis of Binary Sequences", *Journal of the Royal Statistical Society, B*, **20**, 215–242. [5.4]

Cox, D.R., and Hinkley, D.V. (1974), *Theoretical Statistics*, London: Chapman and Hall. [2.3]

Cramer, H. (1938), "Sur un nouveau théorème-limite de la théorie des probabilités," in *Actualités Scientifiques et Industrielles*, Vol.736, Paris: Hermann & Cie. [1.1, 5.3]

Cramer, H. (1946), *Mathematical Methods of Statistics*, Princeton University Press, Princeton, NJ. [5.3]

Cramer, H. (1962), *Random Variables and Probability Distributions*, Cambridge Tracts, 2nd edition. [2.4]

Daniels, H.E. (1954), "Saddlepoint Approximations in Statistics," *Annals of Mathematical Statistics* **25**, 631–650. [1.1, 3.1, 3.3, 5.2, 5.3, 5.4, 7.6]

Daniels, H.E. (1955), Discussion on the paper by Box and Andersen (1955), *Journal of the Royal Statistical Society, B*, 27–28. [1.1, 7.6]

Daniels, H.E. (1956), "The Approximate Distribution of Serial Correlation Coefficients", *Biometrika* **43**, 169–185. [7.6]

Daniels, H.E. (1960), "Approximate Solutions of Green's Type for Univariate Stochastic Processes", *Journal of the Royal Statistical Society, B*, **22**, 376–401. [7.6]

Daniels, H.E. (1980), "Exact Saddlepoint Approximations," *Biometrika* **67**, 53–58. [3.3]

Daniels, H.E. (1982), "The Saddlepoint Approximation for a General Birth Process", *Journal of Applied Probability* **19**, 20–28. [7.6]

Daniels, H.E. (1983), "Saddlepoint Approximations for Estimating Equations," *Biometrika* **70**, 89–96. [3.3, 4.3, 6.2]

Daniels, H.E. (1987), "Uniform Approximations for Tail Probabilities", *International Statistical Review* **55**, 37–48. [3.3, 3.5, 6.2]

David, H.A. (1981), *Order Statistics*, 2nd ed., New York: Wiley. [4.4]

Davison, A.C., and Hinkley, D.V. (1988), "Saddlepoint Approximations in Resampling Methods", *Biometrika* **75**, 417–431. [7.2, 7.6]

De Bruijn, N.G. (1970), *Asymptotic Methods in Analysis*, 3rd ed., Amsterdam: North-Holland. [3.2]

Debye, P. (1909), *Math. Ann* . **67**, 535–558. [3.2]

Dinges, H. (1986a), "Wierner Germs Applied to the Tails of M-estimators", Preprint 352, Sonderforschungsbereich 123, University of Heidelberg. [6.2]

Dinges, H. (1986b), "Asymptotic Normality and Large Deviations", Preprint 377, Sonderforschungsbereich 123, University of Heidelberg. [6.2]

Dinges, H. (1986c), "Theory of Wiener Germs II", Preprint 378, Sonderforschungsbereich 123, University of Heidelberg. [6.2]

Durbin, J. (1980a), "Approximations for Densities of Sufficient Estimators," *Biometrika* **67**, 311–333. [4.2, 4.3, 5.4, 5.5]

Durbin, J. (1980b), "The Approximate Distribution of Partial Serial Correlation Coefficients Calculated From Residuals from Regression on Fourier Series", *Biometrika* *67*, 335–349. [7.6]

Easton, G.S., and Ronchetti, E. (1986), "General Saddlepoint Approximations With Applications to L-statistics", *Journal of the American Statistical Association* **86**, 420–430. [4.3, 4.4]

Edgeworth, F.Y. (1905), "The Law of Error", *Cambridge Philos. Trans* **20**, 36–66 and 113–141. [2.3]

Efron, B. (1981), "Transformation Theory: How Normal is a Family of Distributions?", *Annals of Statistics* **10**, 323–339. [5.2]

Efron, B. (1987), "Better Bootstrap Confidence Intervals", *Journal of the American Statistical Assoc.* **82**, 171–200. [5.2]

Embrechts, P., Jensen, J.L., Maejima, M., and Teugels, J.L. (1985), "Approximations for Compound Poisson and Polya Processes", *Advances in Applied Probability* **17**, 623–637. [7.6]

Esscher, F. (1932),"On the Probability Function in Collective Risk Theory," *Scandinavian Actuarial Journal* **15**, 175–195. [1.1]

Esseen, C.G. (1942), "On the Liapounoff Limit of Error in the Theory of Probability", *Ark. Mat. Astr. och Fys.* **28A**, 19pp. [2.2]

Feller, W. (1971), *An Introduction to Probability and Its Applications*, Vol. 2, New York: Wiley. [1.1, 2.2, 2.4, 3.4, 5.3]

Feuerverger, A. (1989), "On the Empirical Saddlepoint Approximation", *Biometrika* **76**, 457–464. [7.2]

Field, C.A. (1982), "Small Sample Asymptotic Expansions for Multivariate M-estimates," *Annals of Statistics* **10**, 672–689. [4.2, 4.3, 4.5]

Field, C.A. (1985), "Approach to Normality of Mean and M-estimators of Location", *Canadian Journal Statistics* **13**, 201–210. [5.2]

Field, C.A., and Hampel, F.R. (1982), "Small-sample Asymptotic Distributions of M-estimators of Location", *Biometrika* **69**, 29–46. [4.2, 5.2, 5.3]

Field, C.A. and Massam, H. (1987), "A Diagnostic Function for the Saddlepoint Approximations to the Distribution of an M-estimate of Location", Technical Report, Dalhousie University. [3.5, 5.2]

Field, C.A., and Ronchetti, E. (1985), "A Tail Area Influence Function and Its Application to Testing," *Communications in Statistics* C, **4**, 19–41. [7.3]

Field, C.A., and Wu, E. (1987), "Extreme Tail Areas of a Linear Combination of Chi-squares", Technical Report, Dalhousie University. [6.2]

Fisher, R.A. (1934), "Two New Properties of Maximum Likelihood", *Proc. Royal Soc. A* **144**, 285–307. [5.5]

Gamkrelidzt, N.G. (1980), "A Method for Proving the Central Limit Theorem", *Theory of Probability and its Applications* **25**, 609–614. [3.3]

Goldstein, R.B. (1973), "Chi-square Quantiles", *Algorithm 451*, CACM, 451–453. [6.2]

Good, I.J. (1957), "Saddlepoint Methods for the Multinomial Distribution", *Annals of Mathematical Statistics* **28**, 861–881. [7.6]

Good, I.J. (1961), "The Multivariate Saddlepoint Method and Chi-squared for the Multinomial Distribution", *Annals of Mathematical Statistics* **32**, 535–548. [7.6]

Grad, H. (1949), "Note on N-Dimensional Hermite Polynomials", *Communications on Pure and Applied Mathematics* **2**, 325–330. [2.6]

Hall, P. (1983), "Inverting an Edgeworth Expansion", *Annals of Statistics* **11**, 569–576. [2.3, 2.5]

Hall, P. (1988), "Theoretical Comparison of Bootstrap Confidence Intervals", *Annals of Statistics* **16**, 927–953. [6.3]

Hampel, F.R. (1968), "Contributions to the Theory of Robust Estimation," Ph.D. Thesis, University of California, Berkeley. [2.5, 4.3, 7.3]

Hampel, F.R. (1973), "Some Small-sample Asymptotics," in *Proceedings of the Prague Symposium on Asymptotic Statistics*, ed. Hajek, J., Prague: Charles University, 109–126. [1.1, 2.7, 3.3, 5.2, 5.3, 7.3]

Hampel, F.R. (1974), "The Influence Curve and Its Role in Robust Estimation," *Journal of the American Statistical Association* **69**, 383–393. [2.5, 4.3]

Hampel, F.R. (1983), "The Robustness of Some Nonparametric Procedures", in *A Festschrift for Erich L. Lehmann: In Honor of his Sixty-Fifth Birthday*, eds. Bickel, P.J., Doksum, K.A., Hodges, J.L. Jr., Statistics/Probability Series, Belmont (CA): Wadsworth. [7.3]

Hampel, F.R., Ronchetti, E.M., Rousseeuw, P.J., and Stahel, W.A. (1986). *Robust Statistics: the Approach Based on Influence Functions*, New York: Wiley. [2.5, 7.3]

Helmers, R. (1977), "The Order of the Normal Approximation for Linear Combinations of Order Statistics With Smooth Weights Functions", *Annals of Probability* **5**, 940–953. [2.2]

Helmers, R. (1979), "Edgeworth Expansions for Trimmed Linear Combinations of Order Statistics," in *Proceedings of the Second Prague Symposium on Asymptotic Statistics*, eds. Mandl, P. and Huskova, M., Prague: Charles University, 221–232. [2.5, 4.4]

Helmers, R. (1980), "Edgeworth Expansions for Linear Combinations of Order Statistics With Smooth Weight Functions," *Annals of Statistics* **8**, 1361–1374. [2.5, 4.4]

Helmers, R., and van Zwet, W.R. (1982), "The Berry-Esseen Bound for U-statistics", in *Statistical Decision Theory and Related Topics*, eds. Gupta, S.S. and Berger, J.O., Vol. 1, Academic Press, 497–512. [2.2]

Helstrom, C.W. (1978), "Approximate Evaluation of Detection Probabilities in Radar and Optical Communications", *IEEE Transactions on Aerospace and Electronic Systems*, AES-14, **4**, 630–640. [7.5]

Helstrom, C.W. (1979), "Performance Analysis of Optical Receivers by the Saddlepoint Approximation", *IEEE Transactions on Communications*, COM-27, **1**, 186–191. [7.5]

Helstrom, C.W. (1985), "Computation of Photoelectron Counting Distributions by Numerical Contour Integration", *Journal of the Optical Society of America* **2**, 674–6812. [7.5]

Helstrom, C.W. (1986a), "Calculating Error Probabilities for Intersymbol and Cochannel Interference", *IEEE Transactions on Communications*, COM-34, **5**, 430–435. [7.5]

Helstrom, C.W. (1986b), "Significance Probabilities for Wilcoxon Rank Tests Evaluated by Numerical Contour Integration", manuscript. [7.4]

Helstrom, C.W., and Rice, S.O. (1984), "Computation of Counting Distributions Arising from a Single-Stage Multiplicative Process", *Journal of Computational Physics* **54**, 289–324. [7.5]

Helstrom, C.W., and Ritcey, J.A. (1984), "Evaluating Radar Detection Probabilities by Steepest Descent Integration", *IEEE Transactions on Aerospace and Electronic Systems*, AES-20, **5**, 624–634. [7.5]

Helstrom, C.W., and Ritcey, J.A. (1985), "Evaluation of the Noncental F-Distribution by Numerical Contour Integration", *SIAM Journal of Scientific and Statistical Computing* **6**, 505–514. [7.6]

Henrici, P. (1977), *Applied and Computational Complex Analysis*, Vol. 2, New York:

Wiley. [2.1]

Hoeffling, W. (1948), "A Class of Statistics With Asymptotically Normal Distributions", *Annals of Mathematical Statistics* **19**, 293–325. [2.5]

Holly, A. (1986), "Tensor Components of Multivariate Hermite Polynomials and Moments of a Multivariate Normal Distribution", Cahier 8701, Département d'économétrie et d'économie politique, HEC, Université de Lausanne, Switzerland. [2.6, 7.6]

Holly, A., and Phillips, P.C.B. (1979), "A Saddlepoint Approximation to the Distribution of the k-Class Estimator of a Coefficient in a Simultaneous System", *Econometrica* **47**, 1527–1547.[7.6]

Hougaard, P.(1985) , "Saddlepoint Approximations for Curved Exponential Families", *Statistics and Probability Letters* **3**, 161–166. [5.4]

Huber, P.J. (1964), "Robust Estimation of a Location Parameter", *Annals of Mathematical Statistics* **35**, 73–101. [4.2, 4.5]

Huber, P.J. (1965), "A Robust Version of the Probability Ratio Test", *Annals of Mathematical Statistics* **36**, 1753–1758. [7.3]

Huber, P.J. (1967), "The Behaviour of Maximum Likelihood Estimates Under Nonstandard Conditions", *Proc. Fifth Berkeley Symp. Math. Stat. Prob.* **1**, 221–233. [4.2, 7.3]

Huber, P.J. (1981), *Robust Statistics*, New York: Wiley. [4.5, 6.3, 7.3]

Imhof, J.P. (1961), "Computing the Distribution of Quadratic Forms in Normal Variables", *Biometrika* **48**, 419–426. [7.6]

James, G.S., and Mayne, A.J. (1962), "Cumulants of Functions of Random Variables", *Sankhyā Ser. A* **24**, 47–54. [4.2, 4.5]

Jeffreys, H., and Jeffreys, B.S. (1950), *Methods of Mathematical Physics*, Cambridge University Press. [3.3]

Jensen, J.L. (1988), "Uniform Saddlepoint Approximations", *Advances in Applied Probability* **20**, 622–634. [3.3]

Johns, M.V. (1988), "Importance Sampling for Bootstrap Confidence Intervals", *Journal of the American Statistical Association* **83**, 709–714. [7.6]

Keilson, J. (1963), "On the Asymptotic Behaviour of Queues", *Journal of the Royal Statistical Society*, B, **25**, 464–476. [7.6]

Keilson, J. (1965), "The Role of Green's Functions in Congestion Theory" (with discussion), in *Proceedings of the Symposium on Congestion Theory*, eds. W. L. Smith and W.E. Wilkinson, Chapel Hill, NC: University of North Carolina Press, 43–71. [7.6]

Kendall, M., and Stuart, A. (1977), *The Advanced Theory of Statistics*, Vol. 1, Fourth Edition, London: C. Griffin & Company. [2.3]

Khinchin, A.I. (1949), *Mathematical Foundations of Statistical Mechanics*, New York: Dover Publications. [1.1, 3.4]

Kullback, S. (1960), *Information Theory and Statistics*, New York: Wiley. [3.4]

Lambert, D. (1981), "Influence Functions for Testing, *Journal of the American Statistical Association* **76**, 649–657. [7.3]

Lugannani, R., and Rice, S. (1980), "Saddle Point Approximation for the Distribution of the Sum of Independent Random Variables. *Advances in Applied Probability* **12**, 475–490. [3.5, 4.2, 5.2, 6.1, 6.2, 7.6]

Lugannani, R., and Rice, S.O.(1984), "Distribution of the Ratio of Quadratic Forms in Normal Variables - Numerical Methods", *SIAM Journal of Scientific and Statistical Computing* **5**, 476–488. [7.6]

McCullagh, P. (1984), "Local Sufficiency", *Biometrika* **71**, 233–244. [5.5]

McCullagh, P. (1987), *Tensor Methods in Statistics*, New York: Chapman and Hall. [2.6, 5.2]

Parr, W., and Schucany, W. (1982), "Jackknifing L-statistics With Smooth Weight Functions," *Journal of the American Statistical Association* **27**, 639–646. [4.4]

Petrov, V.V. (1965), "On the Probabilities of Large Deviations for Sums of Independent Random Variables", *Theory of Probability Applications* **10**, 287–298. [7.6]

Phillips, P.C.B. (1978), "Edgeworth and Saddlepoint Approximations in a First Order Autoregression", *Biometrika* **65**, 91–98. [7.6]

Pollard, D. (1985), "New Ways to Prove Central Limit Theorems", *Econometric Theory* **1**, 295–314. [1.1, 2.2]

Reid, N. (1988), "Saddlepoint Methods and Statistical Inference", *Statistical Science* **3**, 213–238. [1.2, 5.4, 5.5]

Renyi, A. (1959), "On Measures of Dependence", *Acta Math. Acad. Sci. Hung.* **10**, 441–451. [4.5]

Rice, S.O. (1973), "Efficient Evaluation of Integrals of Analytic Functions by the Trapezoidal Rule", *Bell System Technical Journal* **52**, 707–722. [7.5]

Richter,W. (1957), "Local Limit Theorems for Large Deviations", *Theory of Probability Applications* **2**, 206–220. [5.3]

Rieder, H. (1980). "Estimates Derived from Robust Tests. *Annals of Statistics* **8**, 106–115. [7.3]

Riemann, B. (1892), *Riemann's Gesammelte Mathematische Werke*, New York: Dover Press, (1953), 424–430. [3.2]

Ritcey, J.A. (1985), "Calculating Radar Detection Probabilities by Contour Integration", Ph.D. Thesis, University of California, San Diego. [2.2, 7.5]

Robinson, J. (1982), "Saddlepoint Approximations for Permutation Tests and Confidence Intervals", *Journal of the Royal Statistical Society*, B, **44**, 91–101. [6.2, 7.6]

Robinson, J., Höglund, T., Holst, L., and Quine, M.P. (1988), "On Approximating Probabilities for Small and Large Deviations in $\mathbf{R}^d$", University of Sydney, manuscript. [7.4]

Ronchetti, E. (1979), "Robustheitseigenschaften von Tests", Dipl. Thesis, ETH Zürich. [7.3]

Ronchetti, E. (1982). Robust Testing in Linear Models: the Infinitesimal Approach", Ph.D. Thesis, ETH Zürich. [7.3]

Ronchetti, E. (1987), "Robust C($\alpha$)-Type Tests for Linear Models", *Sankhyā* **49**, Series A, 1–16. [7.3]

Ronchetti, E. (1989), "Density Estimation Via Small Sample Asymptotics", manuscript. [7.2]

Ronchetti, E., and Welsh, A. (1990), "Empirical Small Sample Asymptotics", manuscript. [7.2]

Rousseeuw, P.J., and Ronchetti, E. (1979), "The Influence Curve for Tests", Research Report 21, Fachgruppe für Statistik, ETH Zürich. [7.3]

Rousseeuw, P.J., and Ronchetti, E. (1981), "Influence Curves for General Statistics", *Journal of Computational and Applied Mathematics* **7**, 162–166. [7.3]

Serfling, R. (1980), *Approximation Theorems of Mathematical Statistics*, New York: Wiley. [1.1, 2.2]

Shorack, G.R. (1969), "Asymptotic Normality of Linear Combinations of Functions of Order Statistics", *Annals of Mathematical Statistics* **40**, 2041–2050. [4.4]

Shorack, G.R. (1972), "Functions of Order Statistics," *Annals of Mathematical Statistics* **43**, 412–427. [4.4]

Skovgaard, I.M. (1985), "Large Deviation Approximations for Maximum Likelihood Estimates", *Prob. Math. Statist* **6**, 89–107. [5.4]

Skovgaard, I.M. (1986), "On Multivariate Edgeworth Expansions", *International Statistical Review* **54**, 169–186. [2.5, 2.6]

Skovgaard, I.M.(1987), "Saddlepoint Expansion for Conditional Distributions", *Journal of Applied Probability* **24**, 875–887. [5.5]

Spady, R.H. (1987), "Saddlepoint Approximations for Regression Models", Technical Report, Bell Communications Research, Morristown NJ. [4.5]

Stigler, S.M. (1969), "Linear Functions of Order Statistics", *Annals of Mathematical Statistics* **40**, 770–788. [4.4]

Stigler, S.M. (1974), "Linear Functions of Order Statistics With Smooth Weight Functions," *Annnals of Statistics* **2**, 676–693. [4.4]

Tierney, L. (1990), *LISP-STAT: Statistical Computing and Dynamic Graphics in LISP*, New York: Wiley (to appear). [7.6]

Tierney, L., and Kadane, J.B. (1986), "Accurate Approximations for Posterior Moments and Marginal Densities", *Journal of the American Statistical Association* **81**, 82–86. [7.6]

Tingley, M.A. (1987), "The Lugannani-Rice Tail Area Approximation With Application to the Bootstrap", Technical Report, Dalhousie University. [3.5, 4.2, 6.2]

Tingley, M.A. and Field, C.A. (1990), "Small sample Confidence Intervals", *Journal of American Statistical Association* (to appear).

Tukey, J.W. (1977), *Exploratory Data Analysis*, Reading (MA): Addison-Wesley. [4.4]

van Zwet, W.R. (1979), "The Edgeworth Expansion for Linear Combinations of Uniform Order Statistics," in *Proceedings of the Second Prague Symposium on Asymptotic Statistics*," eds. Mandl, P. and Huskova, M., Prague: Charles University, 93–101. [2.5, 4.4]

van Zwet, W.R. (1984), "A Berry-Esseen Bound For Symmetric Statistics", *Zeitschrift für W'keits-theorie und Verwandte Gebiete* **66**, 425–440. [2.2]

Visek, J.A. (1983), "On Second Order Efficiency of a Robust Test and Approximations of its Error Probabilities", *Kybernetika* **19**, 387–407. [7.3]

Visek, J.A. (1986), "A Note on Numerical Aspects of Robust Testing", *Problems of Control and Information Theory* **15**, 299–307. [7.3]

von Mises, R. (1947), "On the Asymptotic Distribution of Differentiable Statistical Functions," *Annals of Mathematical Statistics* **18**, 309–348. [2.5, 4.3]

Wallace, D.L. (1958), "Asymptotic Approximations to Distributions", *Annals of Mathematical Statistics* **29**, 635–654. [2.3]

Watson, G.N. (1948), *Theory of Bessel Functions*, Cambridge University Press. [3.3]

Weisberg, H. (1971), "The Distribution of Linear Combinations of Order Statistics From the Uniform Distribution", *Annals of Mathematical Statistics* **42**, 704–709. [4.4]

# AUTHOR INDEX

# SUBJECT INDEX